THE WORLD
世界建筑事务所精粹
ARCHITECTURAL III
FIRM SELECTION

COMMERCIAL BUILDING / OFFICE BUILDING /
COMPLEX BUILDING / CULTURAL BUILDING

商业建筑 / 办公建筑 / 综合建筑 / 文化建筑

深圳市博远空间文化发展有限公司 编

天津大学出版社
TIANJIN UNIVERSITY PRESS

PREFACE

PREFACE

In the past few decades, the practice of architectural design has developed towards the postmodern design concept with collage, mixture and coexistence from the modernism on the basis of functionalism and single typology. Rationality turns to be the theme of the development clue of architecture from the modernism stage dominated by the geometric form to the postmodernism stage with multiple mixing symbols. Both the architecture and the social activities of human penetrate and promote each other. The history, economics, philosophy, sociology, etc. have exerted unprecedented influence on the theory and practice of architecture design, which enable the contemporary buildings to obtain the ecological characteristics with multi-elements. The thoughts of architecture such as the phenomenology, semiology, structuralism and deconstruction endow the architecture in this era with symbolic interpretations and create the works for this era.

The innovative and pioneering architects open up new possibilities of the functions, space and forms of the architecture with nontraditional innovative spirit and creative practices. The design methodologies based on the information technology, such as the nonlinear design, parametric design and virtual construction, greatly enrich the design measures of the architects, and open the gate towards the architecture in the new era. The unrealized design creations such as non-standardization, structural skin, free form and surreal characteristics under traditional design conditions have displayed themselves with the development of the digital technology, information technology, structure and material technology. The architecture breaks through the restraints from the structure and function, and the shape and space tend to be more complicated and intriguing, displaying the unique characteristics and times imprinting of the contemporary architecture. Extensive excellent cases and architectural practices are provided in this book, which offers architectural cases with different functions, scales, fields and cultural backgrounds. Readers can spy into the influences and expressions of the above the mentioned architectural thoughts and design methodology in the excellent architectural creations, thus achieving thought-provoking inspiration and apperception.

Wang He
January 28, 2013

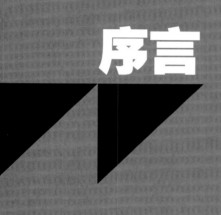

序言

在过去的几十年，建筑设计的实践从功能主义和单一类型学基础上的现代主义向后现代的拼贴、混合、共存并置的设计思想发展。建筑从强调几何形体构成的现代主义时期，到多元符号混合的后现代主义时期，理性成为这一发展线索的主题。建筑与人类的社会活动相互渗透、相互促进。历史学、经济学、哲学、社会学等都对建筑设计理论与实践产生了空前的影响，使当代建筑呈现出多元并存的生态特征；现象学、符号学、结构主义、解构主义等建筑思潮都给这个时代的建筑赋予了标志性的注释，并创造出属于这个时代的作品。

勇于创新和开拓的建筑师以颠覆传统的创新精神和创作实践，开拓了建筑功能、空间、形式的新可能。非线性设计、参数化设计、虚拟建造等后工业时代以信息技术为基础的设计方法论极大地丰富了建筑师的设计手段，打开了通向新时代建筑的大门。非标准化、结构性表皮、自由形体、超现实的特征等在传统设计条件下难以实现的设计创作，随着数字技术、信息技术、结构与材料技术的发展，一一呈现在世人面前，建筑自身突破了结构与功能的约束，造型与空间更趋于复杂与耐人寻味，表现出当代建筑独有的特征和时代印记。

本书提供了大量的优秀范例和建筑实践，呈现了不同功能、不同规模、不同地域和文化背景的建筑案例。读者可以从中窥探到以上建筑思潮和设计方法论在优秀建筑创作中的影响与展现，并从中得到思考性的启发与感悟。

王禾

2013 年 1 月 28 日

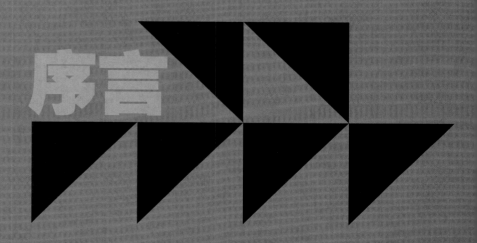

目录
CONTENTS

商业建筑 COMMERCIAL BUILDING

In the narrow sense, commercial buildings refer to public buildings used for exchange and circulation of commodities, as well as places where people are engaged in various business activities. Commercial buildings can be divided into retail stores where all kinds of daily necessities and means of production are sold, shopping malls and wholesale markets, trading places for financial securities, various service buildings such as hotels, restaurants, cultural entertainment facilities, shopping centers, commercial streets and clubs.

Modern commercial buildings emphasize not only the pursuit of commercial interest, but also social value, public interest, cultural taste and the impact on people' s lifestyles. To estimate whether a commercial building is provided with aesthetic value, we not only see its external factors such as decoration materials, facade design, proportion and color, but also emphasize the participation sense of people. Modern society is a people-oriented society, and people are the main body of the society and space. To place people in an architectural building, the scale of the building should be considered carefully. Commercial buildings, especially large commercial building complex, should be complicated but not messy in form, bulky but not distorted, dignified but approachable, and its facade is also a symbol of commercial culture. The commercialization of architecture promotes the diversification of design approach in commercial buildings. After architects' endless quests for innovation, current architectural design methods are gradually maturing. Plenty of new building terms are generated to enrich people' s lives and meet people' s physical and spiritual demands.

Modern commercial buildings play an important role in urban public buildings, impacting the development,construction, image and environment of cities. They also play an important role in modern landscape, containing rich contents in people' s daily lives, enriching the splendor of modern buildings through their unconventional, stylish and beautiful exterior designs. Thus modern commercial buildings have become the iconic symbol of modern civilization.

　　狭义上的商业建筑是指用来进行商品交换和商品流通的公共建筑，以及供人们从事各类经营活动的建筑物。商业建筑的主要类型有销售各类日常用品和生产资料等的零售店、商场和批发市场，金融证券等行业的交易场所，各类服务业建筑如旅馆、餐馆、文化娱乐设施、购物中心、商业街和会所等。

　　现代商业建筑在追求商业利益的同时，也非常重视其自身的社会价值、公共利益、文化品位以及对人们生活模式带来的影响。判断一座商业建筑是否具有审美价值，不仅仅是看它的装修材料、立面设计、比例、色彩等外部因素，更重要的是强调人的参与意识。现代社会是一个以人为本的社会，人是社会的主体，也是空间的主体。建筑形体的尺度经过周密考虑，以使人能置身其中。商业建筑，尤其是大型商业综合体建筑的形体应该繁复而不凌乱，体量庞大而不失真，它既是超人的，又是亲人的，其外立面也是商业文化的表征。建筑的商业化促使了商业建筑外观设计手法的多样化，各式各样的设计五花八门、层出不穷。经过设计师们的努力创新，当下的建筑设计手法正在逐步走向成熟，并且产生了很多建筑类的新词汇，更好地丰富了人们的生活，满足了人们物质和精神的需要。

　　现代商业建筑是城市中的重要公共建筑，它对城市的开发和建设，对城市面貌和环境的塑造产生了重要影响，在现代景观中具有显著的地位，涵盖了人们日常所需的丰富内容。商业建筑标新立异、时尚美观的外形设计为现代建筑增添了无限风采，成为现代文明的标志性符号。

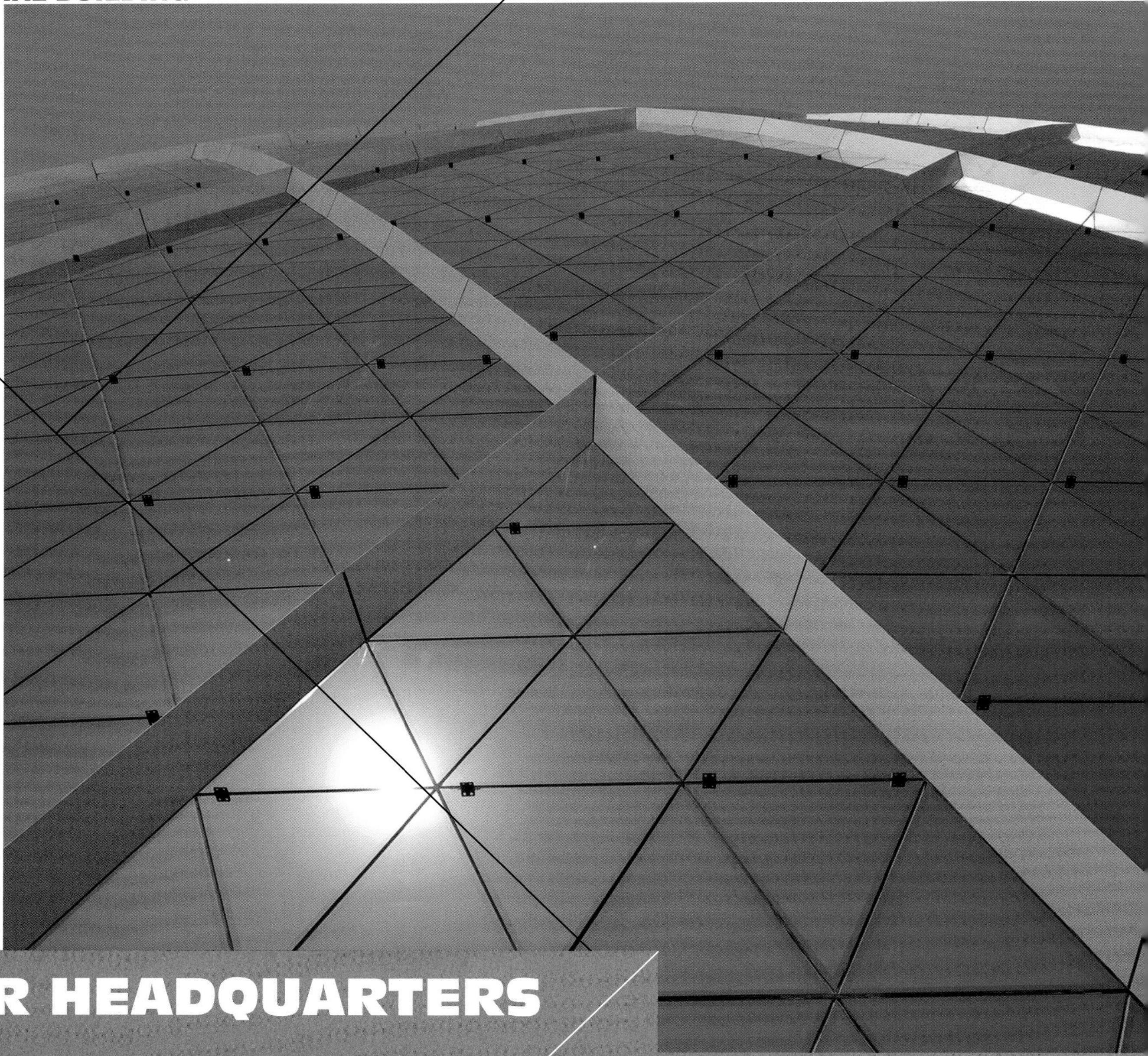

ALDAR HEADQUARTERS
Aldar 总部

Architects: MZ Architects
Location: Abu Dhabi, UAE
Area: 123,000 m^2

设计机构：MZ 建筑设计事务所
项目地点：阿联酋阿布扎比市
面积：123 000 平方米

The idea

Driving along the desert road and with the anticipation of reaching Abu Dhabi a shining element captures one's eyes from a distance. As one moves closer to the object, it becomes obvious that the extensive blurring of scale generally experienced in the desert landscape is once again apparent but this time through an iconic man- made creation: the Headpuarters. In the wake of the construction boom of Abu Dhabi and in an effort to put the area on the map, MZ Architects were commissioned the design of the Aldar Headpuarters, a building that will change and expand Abu Dhabi' s skyline forever. This architectural icon was to shine at the center of Al Raha Beach Development, also a project by Aldar and a newly envisioned microcosm that would bring new life and activity to the Abu Dhabi waterfront. The Headpuarters was to rise from the sea and become a landmark for the area, the city as well as the group behind its creation. For the architect Marwan Zgheib, the power of the monument and the icon lies in its simplicity. His enthusiasm for the project led him to a clear objective: create a simple, daring and powerfully present object that was able to compete with the iconic architecture of the UAE and create a sense of place and identity for the area.

The skin

Just like a seashell, HQ's morphology merged the idea of shape, sculpture and pattern into one unified and expressive whole. The curved glass skin became one of its most complex components to be executed in record time.

The HQ structure has a 25m cantilevered design in every longitudinal direction, concrete couldn't be considered as construction material. A concrete structure would have required unwanted internal supports and entailed time-consuming and costly on-site construction work. Therefore, the team developed a complex external diamond-shaped steel structure called a diagrid, which achieved the striking shape of the building. The first of its kind in the UAE, the diagrid allowed the creation of structural efficiency and stability appropriate to the circular building. The system not only helped minimize the impact of the steel frame on the facade but also served as an architectural element that blurred all sense of scale and inflated the structure, moving away from the typical horizontal stratification of the facades. This diagrid system eliminates the need for internal columns to support the building, which would compromise the aesthetic appeal as well as the views from within. It improves the building's efficiency, providing layout flexibility for tenants.

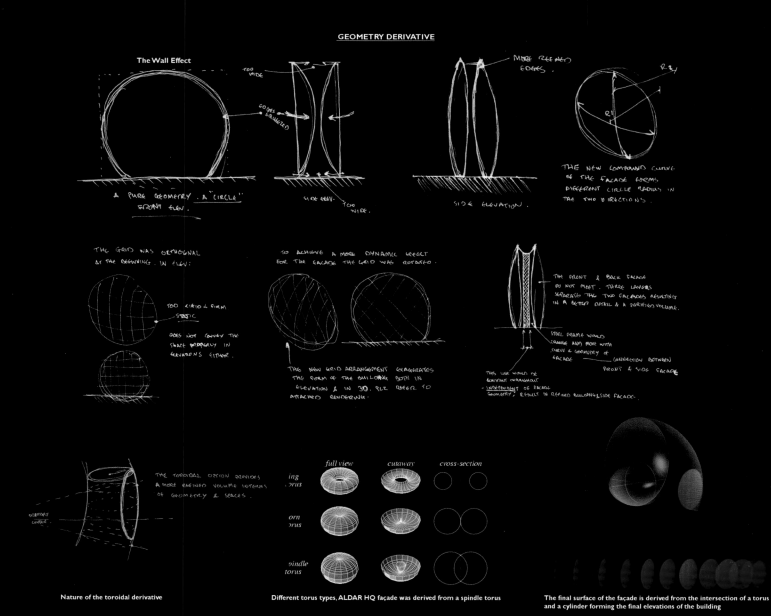

GEOMETRY DERIVATIVE

Nature of the toroidal derivative

Different torus types, ALDAR HQ façade was derived from a spindle torus

The final surface of the façade is derived from the intersection of a torus and a cylinder forming the final elevations of the building

Geometry Derivative　几何导数图

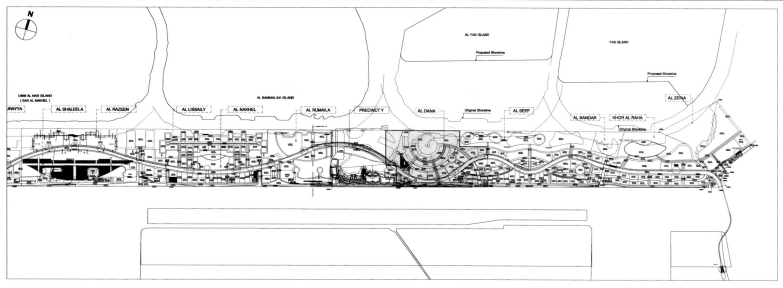

AL RAHA Beach Development–Master Plan   AL RAHA 海滩发展区总平面图

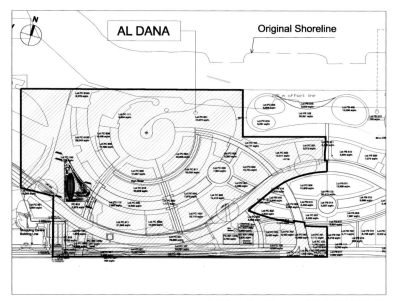

AL DANA–Key Plan   AL DANA 标准层平面图

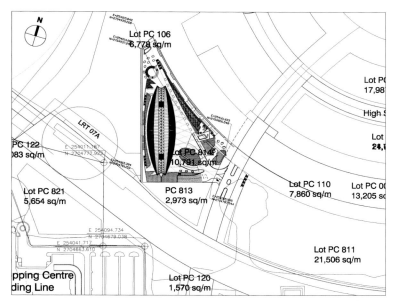

ALDAR HQ–Site Plan   ALDAR HQ 总平面图

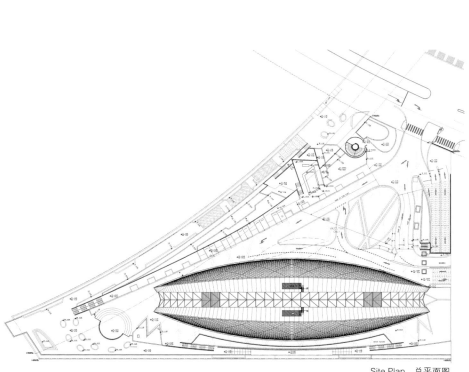

Site Plan   总平面图

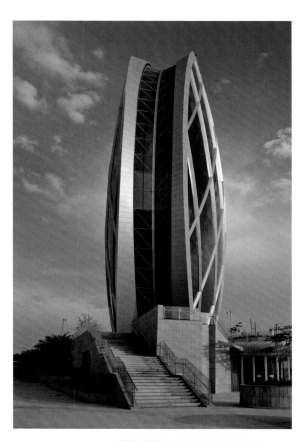

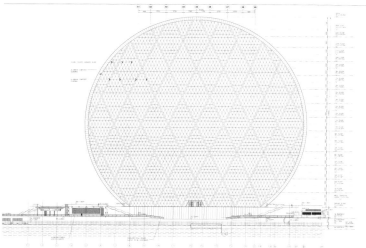

West Elevation　西立面图

构思

当人们沿着沙漠公路驾驶即将到达阿布扎比时，在很远的地方目光就会被一个闪闪发亮的元素吸引。当人们离这座建筑越来越近时，它的形象从抽象模糊的沙漠景观中再次清晰起来，这是一座标志性的总部建筑。在阿布扎比的建筑热潮中，为了将这个地区永远地留在地图上，MZ 建筑设计事务所设计的 Aldar 总部改变和扩展了阿布扎比的天际线。这座标志性建筑在 Al Raha 海滩发展区闪耀光芒，而且为 Aldar 公司服务，由此引入新的微观世界，为阿布扎比市的海滨带来新的生机和活力。本建筑从海洋中升起，将成为地区、城市乃至创造它的团队的一座标志性建筑。对于建筑师 Marwan Zgheib 来说，建筑永恒的和地标性的力量来自建筑的简明直接，他对于这个项目的热情引导他有一个明确的目标：创造一座简单、大胆、有力的建筑，使其能够和阿联酋的任何标志性建筑竞争，并且创造出本地区绝对意义上的地标。

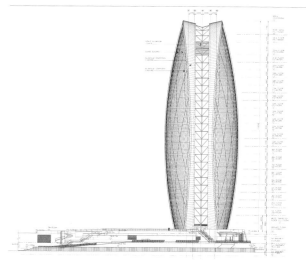

North Elevation　南立面图

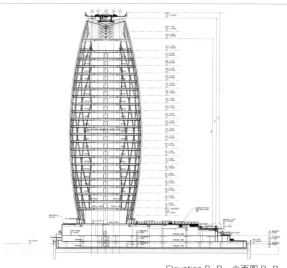

Elevation B–B　立面图 B–B

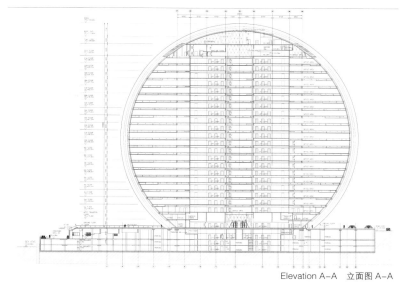

Elevation A–A　立面图 A–A

建筑

建筑的灵感来自蛤壳，蛤壳对于阿布扎比的航海文化具有重大意义。建筑的外形是象征性的几何圆形，建筑师将其设计为两面巨大的圆曲线玻璃墙，就像一个张开的蛤壳。

通过这种构思，一个纯几何形状的大胆设计诞生了：一座具有曲线玻璃幕墙的圆形摩天大楼。这座球形摩天大楼在海域和陆地上恣意炫耀它的外形，就好像沙滩上的一粒珍珠。从远处就可以看到该建筑，无论距离远近它都给人留下深刻印象。它的永恒的几何外形和完美深远的象征意义极大地丰富了建筑的表现力，但是也给结构的稳定性带来极大的挑战。建筑从简单、纯粹和向大自然学习的理念中发展而来，在设计时和最古老的建筑规则——比例相结合。事实上，为了帮助建筑师实现其想法，解决圆形建筑的稳定性是至关重要的。建筑师们接受挑战，回溯了宇宙中的维特鲁威人人体圆环，将五角星嵌入建筑的圆形外立面，可以找到使其稳定的两个点，这两个点就是建筑的着地点，这就是建筑师完成这个挑战的方法。许多简单的建筑都具有复杂性，Aldar 总部就是这样的一个例子，它融合了建筑和结构。像许多大自然的鬼斧神工一样，本建筑的每一个单体在规定时间内分别完成然后形成一个整体。

表皮

Aldar 总部的建筑结构在每一个纵向方向上都有一个 25 米的悬臂结构，混凝土由于需要很多不必要的内部支撑、养护时间漫长以及现浇制作耗资巨大而不能被选做建筑材料，因此，建设团队开发了一种复杂的菱形钢结构，这种被称为斜肋构架的结构组成了建筑的惊人外形。斜肋构架首次被用于阿联酋，它能够形成适用于圆形建筑的结构强度和稳定性。这个系统不仅可以帮助减少钢结构框架对于建筑外立面的影响，而且建筑元素可以模糊建筑的尺寸效应，扩大建筑的视觉范围和摒弃传统建筑层的外立面。斜肋框架系统并不需要内部框架来支撑建筑结构，这将保持建筑外部和内部的美感，提高建筑的效率，并为承租人提供布局灵活性。

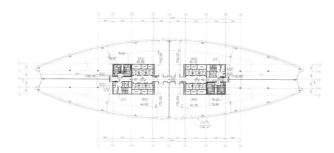

11th Floor Plan　11 楼平面图

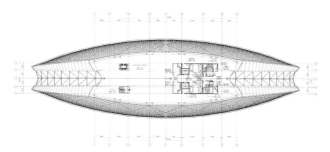

23th Floor Plan　23 楼平面图

Ground Floor Plan　底层平面图

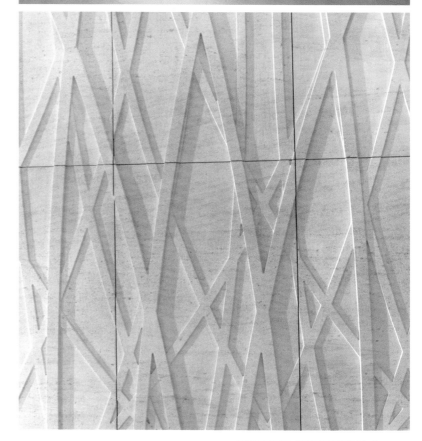

THE CENTAURUS
ISLAMABAD, PAKISTAN
伊斯兰堡 The Centaurus 综合体

Architects: Atkins
Location: Islamabad, Pakistan
Area: 28,125 m^2
Client: Pak Gulf Construction Pvt. Ltd.

设计机构：Atkins 公司
项目地点：巴基斯坦伊斯兰堡市
面积：28 125 平方米
客户：巴基斯坦海湾建设有限公司

The whole of the project Centauras development consists of 7-star Hotel, corporate complex, two residential towers and international standard shopping mall by Pak Gulf Constructions Pvt. Ltd., a prominent real estate builder based in Karachi. Islamabad epitomizes the ambitions of a young and dynamic city that has set its eyes on a magnificent future. Nestled in the commercial heart of the capital on the confluence of the main Faisal & Jinnah Avenues in sector F-8 lies the prime corner site of the Centaurus. The Centaurus consists of 28,125 m^2 of architectural brilliance in the heart of Islamabad. It can bear up to 9.5 magnitude earthquake and was completed in 2010.

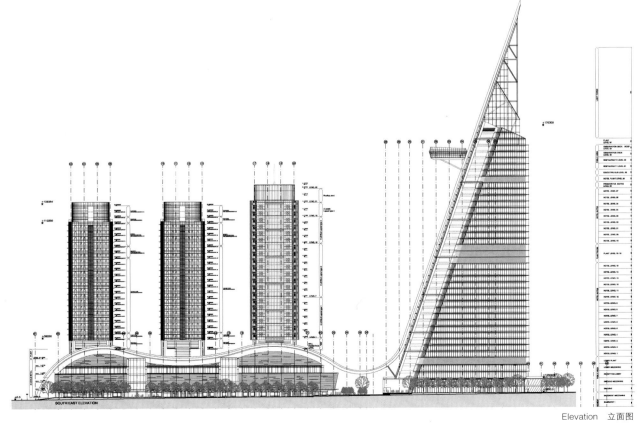

Elevation 立面图

整个 Centaurus 项目由七星级酒店、企业综合体、两栋住宅大楼以及国际化标准购物中心组成，项目由总部设在卡拉奇的知名房地产公司巴基斯坦海湾建设有限公司完成。伊斯兰堡这个年轻动感且富有野心的城市将它的目光放在了长远的未来。本建筑坐落在首都的商业中心，位于 F8 区繁忙的主街 Faisal 街和 Jinnah 街的交会处。这座占地面积 28 125 平方米的建筑将成为伊斯兰堡市闪耀的中心，它可以承受震级高达 9.5 级的地震，主体建筑于 2010 年完成。

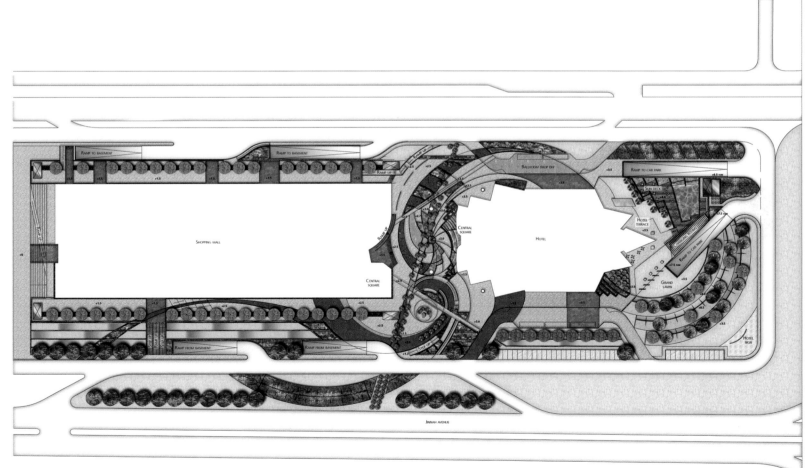

Masterplan　规划图

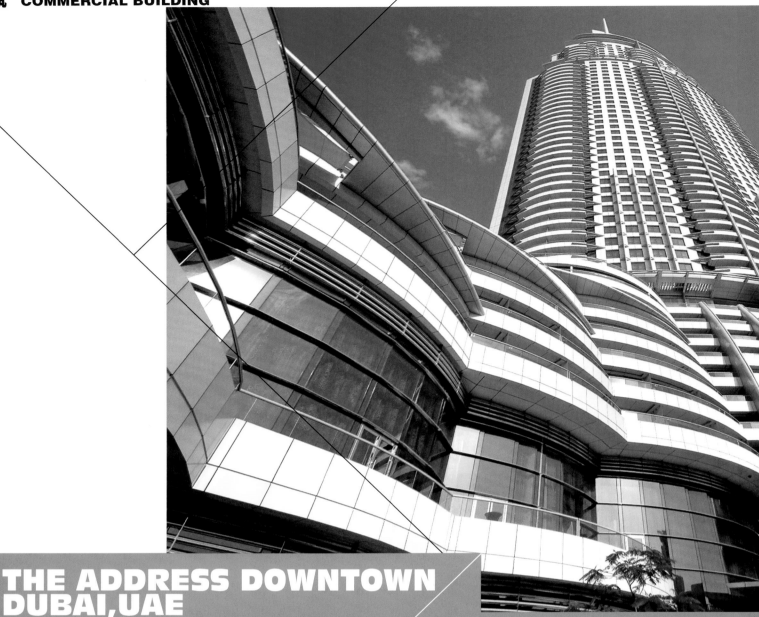

THE ADDRESS DOWNTOWN DUBAI,UAE
阿联酋迪拜 The Address 酒店

Architects: Atkins
Location: Dubai, UAE
Area:178,000 m²

设计机构：Atkins 公司
项目地点：阿联酋迪拜
面积：178 000 平方米

Atkins was responsible for the architectural and engineering design of this five-star hotel and serviced apartment building located in the prestigious Business Bay area. The Address is a mixed-use, 63-storey tower comprising a 198-key, 5-star hotel plus 626 serviced apartments. The imposing 306 m height and striking design produce a landmark project that stands at the opposite end of the lake to the world's tallest building, the Burj Khalifa. The hotel also features a business centre, health club and spa, a restaurant and parking for up to 900 vehicles.

The floor plate of the hotel is based on an organic evolution of a series of linked arcs which present imposing interconnected double and triple volume public spaces with full views of the lake and Burj Khalifa tower; this concept is also extended to the guest room levels. Rising above the hotel is an aerofoil-shaped tower that accommodates 45 levels of deluxe serviced apartments.

Our Carbon Critical Design™ ethos ensured features such as air conditioning condensation recovery process that captures enough water to service the irrigation needs of the whole building.

Facilities provided in the building include:
• commissionary kitchen and extensive back of house facilities
• central laundry which has the capacity to service not only the needs of the 5-star hotel but also four other hotels in the surrounding area
• an all day dining restaurant
• 1,100 m² swimming pool with pool bar
• ballroom and support facilities
• club lounge

The now complete hotel supports the needs of up to 4,000 people who work, live and stay in the building every day.

In meeting the ambitious design programme and providing innovative cost solutions along the way, Atkins built a strong relationship and moved on to work with the client on other projects.

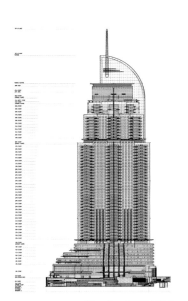

Elevation 1 立面图 1

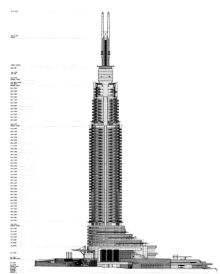

Elevation 2 立面图 2

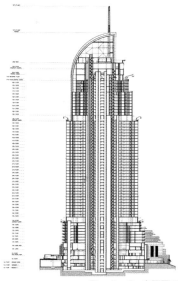

Elevation 3 立面图 3

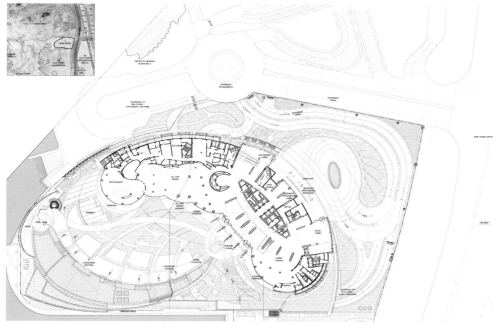

Plan 1　平面图 1

Address 酒店及酒店式公寓位于迪拜商业中心，Atkins 公司负责这座五星级酒店的建筑设计和工程设计。

高达 306 米的 Address 酒店婷婷玉立于湖边，与迪拜塔摩天大楼（Burj Khalifa）隔湖相望。这座地标建筑是一个 63 层的五星级奢华综合体，配套 198 间客房和 626 间酒店式公寓套房，还配套了商业中心、健身中心、水疗中心、餐厅和一个有 900 个车位的超大型停车场，为入住者提供充分的便利。

酒店的楼面形式由一系列相连接的圆弧演变而成，从而形成一系列相连接的两层／三层通高的公共空间，使酒店成为观赏湖景和 Burj Khalifa 大厦的绝佳地点，这一设计理念同时延伸到酒店客房设计中，使得入住的客人享受最佳的湖光美景。位于酒店上部的是机翼形状的 45 层豪华酒店式公寓。

在整个项目设计中，我们运用低碳设计理念，如采用空调冷凝回收系统作为水资源的回收利用方式，为整座建筑的灌溉系统提供充沛的水资源。

建筑内设施包括酒店厨房设施、酒店干洗服务（不仅服务于 Address 酒店，还服务于周边其他四家酒店）、餐饮、泳池、宴会厅及相关设施、酒吧。

现在，Address 酒店每天可以接待 4 000 多人，为工作与生活于其中的各类人群提供更多样化的服务。

在整个项目合作中，Atkins 公司为客户提供了优质的服务，尤其是提出创新性成本解决方案，深得客户的肯定，为 Atkins 公司赢得该客户其他项目奠定了基础。

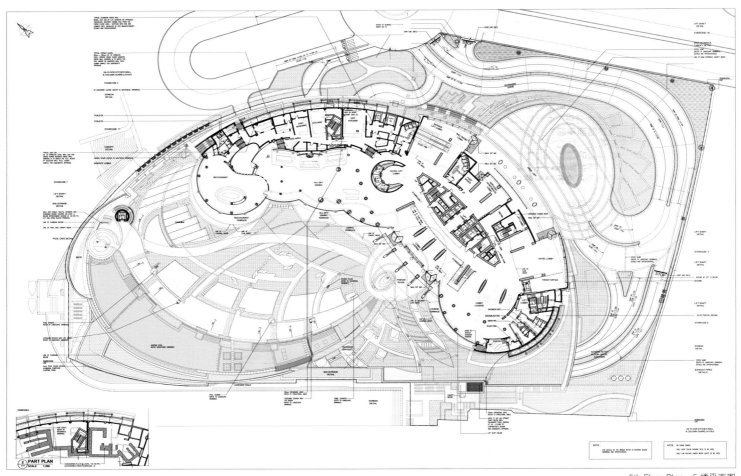

5th Floor Plan　5楼平面图

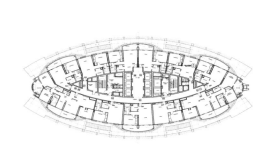

Typical Floor Plan　标准层平面图 1

Typical Floor Plan　标准层平面图 2

RECONSTRUCTION, ADAPTATION AND EXTENSION OF GRAND HOTEL DONAT
度假村的重建、整改和扩张

Architects: API ARHITEKTI
Location: Rogaška Slatina, Slovenia
Area: existing 9,200 m² + extensions 2,600 m²
Photographer: Miran Kambic

设计机构：API ARHITEKTI
项目地点：斯洛文尼罗加斯卡斯拉提纳店
面积：原有 9 200 平方米，另外扩建 2 600 平方米
摄影：Miran Kambic

A comprehensive renovation of Donat hotel, designed by architect Borut Pecenko in 1974, is a part of accomplishing a refreshed image of an established architecture with an aim to carry out programme and functional modernisation. The renovation comprises interventions at various parts of the building. Previously opened terrace at the hotel restaurant is entirely glazed, designed as a winter garden with glass roof supported by up to 16-metre long glass beams. The expressive architecture opens up towards the park and represents a counterpoint to the intimate space of the restaurant. In correspondence to new winter garden, on a flat roof of inner pool, a glass pavillion for relaxation is constructed, as a part of adapted and extensioned wellness, fitness and sauna programme. On ground level, next to adapted inner pool, an outer pool with sunny terrace is built. The renovation of hotel is emphasised with superstruction of luxury-room penthouse on top of the building, and overall new interior design, especially of entrance lobby.

赌场度假村由建筑师 Borut Pecenko 于 1974 年设计，为了执行计划和实现功能现代化，对其全面整修是要更新历史悠久的建筑形象。整修涉及建筑的各个部分。刚刚开放的酒店餐厅的露天平台给人眼前一亮的感觉——被设计成一个由 16 米长的玻璃梁支撑的有着玻璃屋顶的温室。富于表现力的建筑面向公园，并与餐厅私密空间相对照。与新的温室相适应，还建造了内部带屋顶平台的游泳池以及放松用的玻璃馆，作为整改及扩大健身和桑拿项目的一部分。在地面上，整改过的内部游泳池附近建造了带阳光露台的外部游泳池。酒店的整修重点放在建筑顶部带有豪华房间的顶层公寓以及全新的内部设计，尤其是大厅入口。

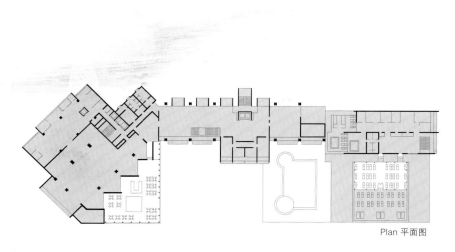

Plan 平面图

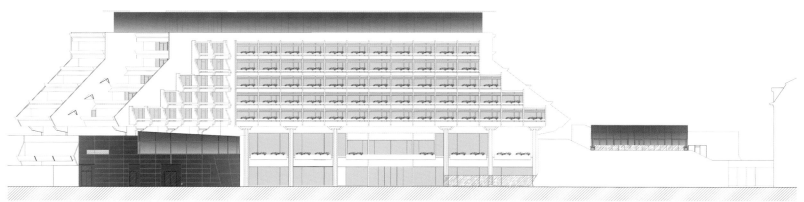

Elevation 立面图

JAKARTA LUXURY APARTMENT, INDONESIA
印度尼西亚雅加达奢华公馆

Architects: Atkins UK
Location: Jakarta, Indonesia
Area: 376,550 m²
Client: Badan Kerjasama Mutiara Buana

设计机构：英国 Atkins 公司
项目地点：印度尼西亚雅加达市
面积：376 550 平方米
客户：Badan Kerjasama Mutiara Buana

The project comprises a 5-star Tower Hotel, luxury category apartment towers, serviced apartments, water park and basement parking facilities developed on 27.15 acres (10.99 hectares) of the northernmost piece of land by the Pantai Mutiara canal estate.

The estate is one of the most luxurious housing complexes in Jakarta and the site is designed for an apartment and hotel development. The project brief was to target the top level of the high-class apartment market and architecturally the development to be a landmark on the Jakarta waterfront. In its design philosophy, Atkins always incorporates the strength of both architecture and engineering to deliver projects that are exceptionally pleasing in form and detail, exceeding our client's expectations.

The concept was to use a nautical theme and orientate the buildings on the cardinal points of the compass thus creating the best possible views of the waterfront for each phase of development. The cluster of apartment buildings represents elegant yachts sailing for distant locations on the cardinal points of the compass. The sailor-shaped architectural fin treatment on the apartment facade also emphasises the nautical character for each apartment cluster. The concept is simple but dynamic and the development is now a landmark on the Pantai Mutiara canal estate when viewed from land, air and sea. The composition is also structured to suit the client's required phasing strategy so that each phase of the project can be developed in stages.

Regatta, Jakarta has been formally acclaimed with a Prix d'Excellence, awarded by the International Real Estate Federation, FIABCI. The Bali Congress Award was awarded by an international panel of top real estate professionals and experts, at a ceremony held during FIABCI's annual world congress held in Bali in 2010.

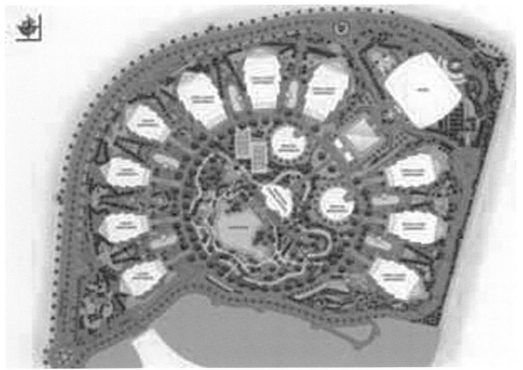

本项目由一座五星级酒店、奢华公寓、酒店式公寓、水景公园和地下停车场组成，项目占地27.15 英亩（10.99 公顷），位于城市最北端，由Pantai Mutiara 运河地产开发。

该地产位于雅加达地区最奢华的住宅区，被设计为一个包含高档公寓和酒店的新社区。本项目针对顶级高端公寓市场而开发，并将会发展成为雅加达海滨地区的地标式建筑。Atkins 公司具有超凡的实力，他们可以同时完成建筑的设计和工程建设，并且极其关注细节，往往能超出客户的期望。

项目以航海概念为主题，使用罗盘为建筑朝向精确定位，从而使建筑群的各个区段都能得到最佳的海景视野。建筑群体现了优雅的游艇航行主题，在很远处就确定了基本的罗盘方位点。帆形建筑的外立面做鱼鳍状处理，从而强调了每个建筑群的航海特点。新开发的楼盘直接简单，富于活力和创新性，已经成为 Pantai Mutiara 运河地产的海陆空新地标。建筑结构化的构图可满足客户的需求，设计师不断调整建设策略，以便建筑每一个阶段的建设都能有条不紊地进行。

在竞赛方面，该项目已经被国际房地产联合会（FIABCI）正式授予优秀奖，并被国际委员会的顶级房地产专业人士和专家授予巴厘岛国会奖，颁奖仪式在 2010 年巴厘岛举行的国际房地产联合会年度世界大会上举行。

Plan 平面图

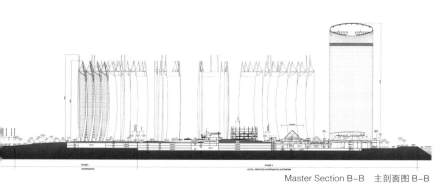

Master Section B-B　主剖面图 B-B

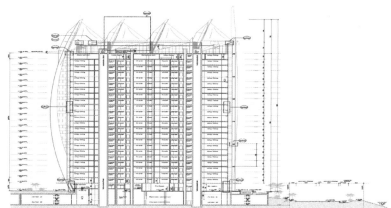

Section B-B　剖面图 B-B

RENOVATION AND EXTENSION OF THE PALACE HOTEL IN PORTOROŽ

波尔托罗日皇宫酒店的装修和扩建

Architects: API ARHITEKTI
Area: 2,804,100 m²
Photographer: Miran Kambic

设计机构：API ARHITEKTI
面积：2 804 100 平方米
摄影：Miran Kambic

Architectural and cultural background of the old Palace Hotel
The old Palace Hotel in Portorož is a landmark that stands as a witness of the development of town and health-wellness tourism on the Slovenian coast. During the last years of the Austro-Hungarian Monarchy guests crowded the halls of the Palace Hotel and enjoyed in prestigious luxury this place offered. The Adriatic Sea was only a stone's throw away from the richly-decorated apartments.
The hotel's design reflects the time in which it was built. The Central Europe's credo of eclecticism is revealed in the building's symmetrical design which is reminiscent of ancient temples erected in the magnificent environment of the Elysian Fields. Idyllic neoclassical architectural composition with its classical design elements gives the observers a sense of safety, firmness, tradition and respect.
Renovation and the bases of design
All characteristics of the existing architecture of the Palace Hotel—a building with a typical oblong ground plan layout; an imposing palace with central symmetry and rich park architecture—were considered during renovation planning. Next to it is situated a newly designed modern building. It coexists in harmony with the existing old building without disturbing the already existing design style. This conveys a message of harmonious coexistence and respect to the same cultural values regardless of the fact that buildings were created in different periods and according to different designs to the observers.

Elevation 立面图

Section 剖面图

The approach to design new architecture of the extension was directed to maintain the original character and message of the old building which served as a base of design. The use of composition, rhythm, materials and colours tries to imitate the strictness of the existing building as closely as possible. This way is, as far as it is possible, placed in a neutral way—nevertheless, it is designed in a modern way—and creates the needed balance with the existing values. Observed from the sea, the combination of stone tiles and greenery of the hotel's facade blends with the hill landscape behind it. Thus the dominant role of the existing facade is kept. Compared to the diversity of design of the surrounding urban environment, it flirts with the restrained design of the facade of the Splošna plovba company building.

From the composition of two standalone buildings which are connected on the ground floor, one can observe basic design idea which is directed at keeping the dominant role of the existing building. The new building's facade design is restrained with clear-cut geometrical composition which is also used in the interior. Thus the extension blends with the surroundings and the old building in an urban entirety with complete design. However, the old palace building keeps its dominant role.

The two buildings are softly connected via a glass entrance hall which receives visitors and offers them a magnificent view of the sea, sky and park. The glass connection between the old and new buildings serves as a see-through filter.

The new northern facade of the existing building and conference centre is clearly distinct from the historicism of the old building. Strict rhythm in which windows are set follows the design of the existing building. In addition, repetitive usage of stone panels is related to the extension design. This way everything is connected in a harmonious whole.

The Building—Water—Park

In front of the Palace Hotel there stands a park situated at the centre with classical symmetrically designed green elements. During renovation a new axis was laid at the intersection of the old and new buildings which runs through the open space in front of the glass connection. The latter connects the north-eastern platform and the sea with a glazed connection. This way the following elements are linked in sequence: the sea, an entrance fountain, a representative stairway with glass curtain and an oval fountain in front of the northern entrance. Moreover, this reflects how the old and new are merged together.

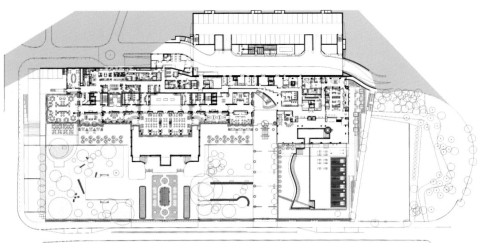

Plan 1 平面图 1

老皇宫酒店的建筑和文化背景

波尔托罗日的老皇宫酒店是见证了斯洛文尼亚海岸线上小镇和健康养生旅游发展的里程碑。在奥匈帝国的最后时期，大量旅客挤满了皇宫酒店的大厅，享受它带来的极致奢华。亚得里亚海距离这座富丽堂皇的建筑仅一步之遥。

该酒店的设计反映了它被建造的那个时代。中欧的折中主义理念在该建筑的对称设计中显现得淋漓尽致，令人想起了矗立在极乐世界中的古老寺庙。融入了经典设计元素的田园新古典主义建筑构图给参观者带来了一种安全、牢固、传统的神圣之感。

翻新和设计基础

在翻新规划中，设计者考虑了现有皇宫酒店建筑的所有特征，即一栋有典型长方形平面图布局的建筑、一座中间对称且有丰富的公园建筑的宏伟宫殿。坐落在它旁边的是一座新设计的现代建筑，它与现有的老酒店和谐共存，未改变原有的设计风格。这给参观者传达了一种和谐共存、对不同文化价值同样尊重的信息，而实际上这些建筑是在不同时期根据不同的设计建造的。

酒店的设计基础在于其扩建方法，它维持了老酒店原有的特点和信息。构造、节奏韵律、材料和色彩的使用方面都尽量模仿现有建筑的严谨性。按这种方式，设计师尽可能将建筑设计成中性风格，但是却用现代风格表达，且用当前的价值观创造出其所需的平衡。从大海方向观察时，石瓦和带有绿植的酒店立面与它后面的丘陵景观融合在一起，因此，现有建筑的主要特色得到保留。与城市周边环境设计的多样性相比，设计师曾一度考虑过采用 Splosna plovba 公司建筑的严谨的立面设计。

从底楼连接起来的两栋独立建筑的构造中，人们可以看到基本的设计思想，即保留现有建筑的主要特色。新建筑的立面设计也由用于室内的清晰的几何构图来控制。因此，酒店的扩建部分与周围的环境融合，老建筑与全新完整的设计在城市中融为了一个整体。但是，老酒店建筑仍保持了其主要特色。

这两座建筑通过玻璃入口大厅连接，玻璃入口大厅接待旅客，给他们呈现了一片壮丽的大海、天空和公园景色。老建筑和新建筑之间的这种玻璃连接起到了透明滤器的作用。

现有建筑和会议中心新的北立面与老建筑的历史主义风格迥然不同。窗户的设置严格遵循原有建筑的设计。此外，石板材的重复使用与扩建设计相关。这样，所有部分形成一个和谐的整体。

建筑—水—公园

皇宫酒店前的中心地段有一座具有经典对称绿色元素设计的公园。在装修过程中，在旧建筑和新建筑的交叉点增加了一条贯穿玻璃大厅前方空地的新轴线。后者通过玻璃大厅与东北角的平台和大海相连。这样，下列元素依次连接：大海、入口处的喷泉、有玻璃幕墙的典型楼梯以及北入口前的椭圆形喷泉。这反映出新旧建筑是如何融为一体的。

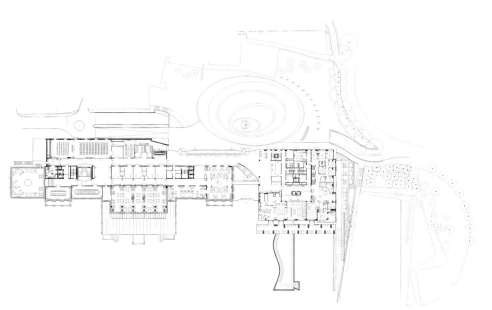

Plan 2 平面图 2

SHANGHAI LINGANG CROWNE PLAZA HOTEL
上海临港新城皇冠假日酒店

Architecs: Atkins, Shanghai
Total area: 69,196 m²
Area above ground level: 50,981 m²
Client: Shanghai Lingang New Town Hotel Investment
Management Ltd Co.

设计机构：Atkins 上海公司
总面积：69 196 平方米
地上面积：50 981 平方米
客户：上海临港新城酒店投资管理有限公司

The Shanghai Lingang Crowne Plaza Hotel was assigned to Atkins Shanghai in the year 2007 through an international architectural competition. After 4 years of design development and construction works, it was completed in 2011 and opened in May, 2012.
Master plan
The hotel is located on a island surrounded by the lake, Which is the city center of Shanghai Lingang New City.Lingang is a satellite town of Shanghai; it is located on the eastern shore of Yangtze delta, 75 km away from the Shanghai city center, 32 km from Yangshan Deep Water Port and 25 km from Pudong Airport.
Right at the heart of the new town is an artificial lake, which occupies an area of 560 hectares. A 16.43 hectares island sits within the circular lake, of which 13.46 hectares of land was allocated for hotel, while the remaining site is reserved for public green area.
The Crowne Plaza Hotel will be one of the pioneer architectural projects in Lingang. Due to its prominent location, the client envisaged the hotel as a signature for the new town. The design team has given the building a distinctive form, like a lotus blossom floating on a reflecting pool. The building was designed to comply with 24 m physical height limit and it is connected to the land via a bridge at south-western corner of the island. The hotel is suited in the north-eastern part of the island for greater exposure to the extensive water surface, while the public green area is at the western tip of the island.

Architecture

There are 356 keys (438 bays) and the hotel is designed as a five-star resort hotel with conference and business facilities. The building took the form of a lotus flower with five petals; the spherical atrium at the center is the hotel lobby, the five wings are guestrooms, restaurants, and meeting rooms.

The petal at the western tip is raised from the ground to facilitate as a drop-off for visitors. Meeting, conference facilities and F&B outlets are located on the ground floor of the southern petals; while the grand ballroom, indoor swimming pool are housed in annexes between the petals.

Due to its unique radial form and unimpeded vista, all hotel rooms are able to enjoy wonderful views and ample daylight. The hotel rooms were planned along single loaded corridors with indoor atriums filled with natural daylight. Room types include deluxe suites, superior rooms, executive rooms, standard rooms and handicapped-access rooms.

The main kitchen is located at the annex between the ballroom and restaurant, while back of the house, service facilities and parking are located at the basement.

With its complex geometry, the building structure has played a major role in the expression of interior design. The structural elements in main atrium and all five petals were clearly revealed in sinuous curves.

The facade design for the hotel is unmistakably unique and modern; the curtain wall systems utilize a rich variety of materials in complex geometry.

The roofs were finished in aluminum standing seam cladding, while insulated glass modules and spandrel panels were applied on the spherical atrium. Terracotta cladding is used extensively on the external wall, and timber panels are applied to partitions for balconies of guest rooms.

Plan 平面图

上海临港新城皇冠假日酒店是 Atkins 上海公司建筑设计部门在 2007 年国际方案竞赛中的优胜方案。经过 4 年的设计、深化以及施工，酒店于 2011 年底宣告落成，并于 2012 年 5 月开业。

总体规划布局

酒店位于一个四面环湖的岛上，这个湖泊是上海临港新城的城市中心。临港新城是上海的卫星城之一，位于长江三角洲东岸，距离上海市中心约 75 公里，南离洋山深水港 32 公里，北距浦东东国际机场 25 公里。

新城的规划遵循霍华德"理想城市"的构思，极富创意地设计了面积达 560 公顷的湖泊作为城市的中心，而 Atkins 公司的设计则是对这个创意的一个贴切的呼应。酒店所在岛屿的面积达 16 .43 公顷，酒店的用地为 13 .46 公顷，其余为公共绿地。

酒店作为新城最先建造的建筑之一，同时也是城市的重要公共建筑和功能核心，业主希望能将酒店打造为一个城市的地标性建筑。我们的设计构思抽象出一种自然的形态，与城市滨海基地环水的地域性呼应：酒店的外形象是莲花漂浮在湖面上，可以唤起人们对海洋的记忆，同时也满足了建筑高度不超过 24 米的规划要求。酒店通过岛上西南侧唯一的桥梁与城市其他部分相连。公共绿地被设置在岛的西侧；酒店则被设置在岛的东北侧，以便最大可能地与湖面相邻，让最多的客房间能够拥有优美的湖景视野。岛中央是大面积的优质果岭草坪，而岛的其余部分被分别设置为泳池区域、运动区域、码头区域等多种娱乐区域，与南岛休闲娱乐运动的定位相契合。

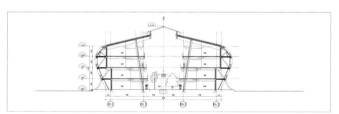

Guest Room Section　客房剖面图

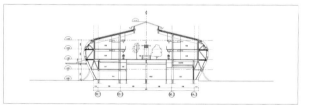

Hotel Section　酒店剖面图

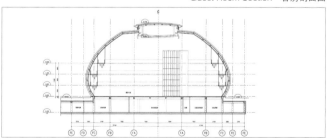

Hotel Lobby Section　酒店大堂剖面图

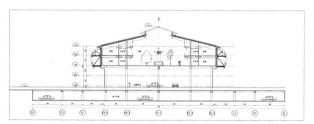

Entrance to the overhead section　入口架空剖面图

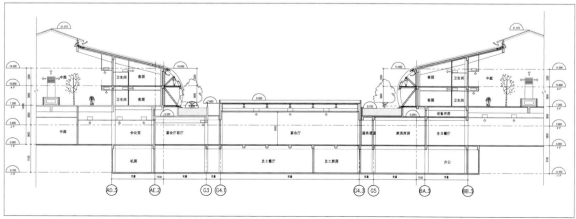

Section　剖面图

建筑单体设计

这是一个拥有356间套房（438间自然间）的五星级商务度假酒店，酒店的功能设置相当完善。酒店的形态如一个花蕾连接着五个花瓣。中心的"花蕾"为酒店四季大厅，是酒店大堂、前台接待及主要交通通道；而"花瓣"部分则被设置为酒店客房、餐厅及会议室等。

酒店西侧靠近中央果岭草地的"花瓣"一、二层架空通高，环岛的主要交通道路从下部通过，是酒店的主入口；酒店南侧"花瓣"的一、二层通高设置为会议区域，共有能容纳40~100人不等的大小会议室8间；酒店东南侧"花瓣"的一、二层通高设置为餐饮区域，包括一个全日餐厅和一个中餐厅。宴会厅位于上述两个"花瓣"中间，便于面积和高度的灵活设置。五个"花瓣"的其余部分均为客房。

基于酒店四周均视线开阔、景观资源非常丰富的特点，由五个伸展的"花瓣"构成的酒店形态使大部分客房都有良好的景观。客房为单侧布置，两侧走廊中间为"花瓣"中庭，中庭顶部的天窗则引入了自然光线和通风。客房类型有酒店套房、豪华套房、行政套房、普通客房、单人间、双人间及无障碍客房。行政套房集中的楼层设有行政酒廊。

主厨房设置在宴会厅和餐厅之间的裙房内。服务区域、大部分的办公区域及厨房库房均设置于地下层，通过楼梯和电梯可直达会议区和主厨房。大多数设备及后勤用房、停车场都设置在地下室。从地下停车场可通过电梯直达酒店大堂。

酒店室内空间设计独具特色。中央的酒店大堂采用全钢结构，螺旋状的钢结构构件形成了优美的弧线，结构外的幕墙龙骨与螺旋状的结构柱如影随形，顶部设有可自动开启的通风百叶。五个"花瓣"部分均设有上下贯通的室内中庭，中庭底层设有绿化庭园，丰富了室内环境。中庭部分每一楼层均设有室内回廊，以作为客房通道。

酒店结构造型独特且现代，复杂的几何立面采用了多种自然材料及先进的幕墙体系，新颖独特而又生态自然。

屋面采用了深色亚光的铝复合板作为外饰材料，使屋面自然延展，创造出优美曲线。大堂为半球形的玻璃体，为降低能耗及创造舒适的室内热工环境，使用了大面积的玻璃影墙和彩釉玻璃。外墙上使用了大量的陶板。客房部分设计了大进深的阳台，采用木质墙面作为阳台之间的隔断。阳台、钢架等元素塑造了虚实相间、层层递进的韵律感。

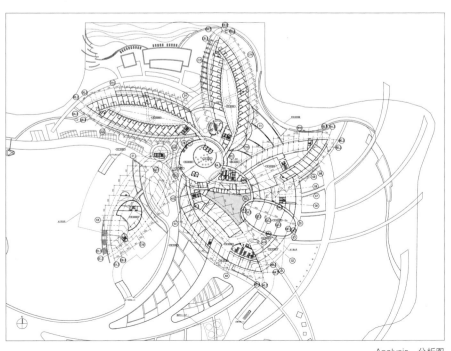

Analysis　分析图

SAMA TOWER
DUBAI, UAE
阿联酋迪拜 Sama 大厦

Architects: Atkins
Location: Dubai, UAE
Area: 93 950 m²
Client: Al Hamid Group

设计机构：Atkins 公司
项目地点：阿联酋迪拜
面积：93 950 平方米
客户：Al Hamid 集团

Situated on the famous Sheikh Zayed Road, the 194 m tall Sama Tower represents a truly modern design concept, providing a unique and dynamic form combined with the most efficient and effective space planning. The theme is contemporary and reflects the vibrant and energetic urban scene of Dubai. Atkins was contracted to carry out the entire design and site supervision services. Challenged by the perpetual debate between form and function, Atkins developed this spirited design in response to the client's aspiration for a unique residential landmark tower in the trade centre area of Dubai. The design of the Sama Tower strives to create the illusion of continuous movement in a static object. This is achieved by the slight twist imparted to the facades that contrive to catch the light and reflect it in a dynamic fashion. More than 700 residential apartments are offered in a mix of one-, two- and three-bedroom apartments, all served by full amenities including a roof-top health club, spa and panoramic gym. A trapezoidal floor plate with a gradually inverting geometry as one goes up represents a twist that is gentle, yet powerful. The reversing triangle also allows for a greater number of apartments to have enhanced views towards the sea on higher floors. The ground and mezzanine floors house boutique retail, food and beverage outlets with a separate 10-storey car park providing exclusive parking for tenants. Eight open plan offices fitted out to "A" grade standard will be provided on each of the nine office floors.

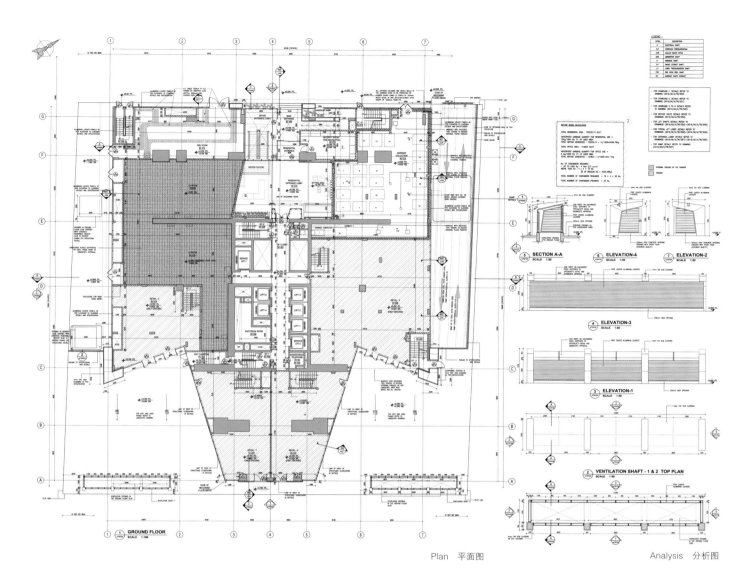

Plan　平面图

Analysis　分析图

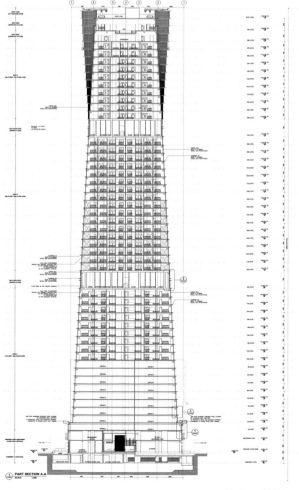

Section　剖面图

高 194 米的 Sama 大厦位于著名的扎耶德街，该大厦体现了真正意义上的现代建筑设计理念，建筑展示了独特动感的形式并具有高效和高利用率的空间规划。建筑的主题很时尚，反映了充满活力的迪拜城市的风光。Atkins 公司完成了本建筑的设计任务并且现场监督了项目的施工。形式和功能的辩证是永久的，Atkins 公司在满足客户希望在迪拜贸易中心区域建设一个唯一的地标性住宅建筑的愿望的基础上做了这个大胆的设计。Sama 大厦努力营造一种在静态对象中产生连续运动幻象的气氛，这凭借建筑外立面的轻微扭曲来完成，使得建筑立面可以捕捉到光，并且反射出动态的效果。700 多套住宅公寓分别为一居室、两居室和三居室，并配有完整的服务设施，如屋顶的健身俱乐部、温泉和全景健身房。建筑配有逐渐升高的反向几何图形的梯形楼板，代表了一种温和而又有力的扭转。建筑外形上倒转的三角形结构可以极大地增加更高楼层拥有海景房间的数量。建筑首层和夹层设有零售商店、食品和饮料的销售点，另有一个独立的 10 层停车场以供专属用户停车。该建筑九个办公楼层均配有八个"A"级标准的开放办公室。

RONGHE PLAZA AND FIVE-STAR HOTEL IN TIANJIN AIRPORT ECONOMIC AREA

天津空港经济区融和广场及五星级酒店

Architects: ZPLUS Architecture and Planning Design Company
Location: Binhai New Area, Tianjin, China
Area: 340,000 m²

设计机构：ZPLUS 普瑞思建筑规划设计公司
项目地点：中国天津市滨海新区
面积：340 000 平方米

Faced with the building mass of over 300,000 m² and the street landscape with the length of more than 600 m, designers have felt a strong sense of social responsibility as well as a sense of mission about how to seize a new urban fragment and make its significant influence stand longer.

Designers are elaborately creating works, graving various themes and presenting different details with an open mind and modern material technique, trying to bring people a rich visual experience and concentrate a new site and time spirit with accumulation and explosion of new materials and new forms.

This modern architecture with the length of 600 m is finally presented before us after elaborate decoration for days and months and modification for 5 years. It took the first place in appraisal for top 10 popular buildings Tianjin Airport Economic Area in 2011.

Project background

Tianjin Airport Economic Area is the west door connecting Beijing and Tianjin in Binhai New Area. After several years of development, five industries including aviation industry, telecommunication industry, equipment manufacture industry, software service outsourcing industry and headquarters economy industry are developing neck to neck here. In the future, characteristics of multiple functions and ecology will be further enhanced, forming gradually high-end business district and high-tech development atmosphere, and it is expected to become "Seattle in the East".

Ronghe Plaza is located in the main entrance of Tianjin Airport Economic Area, adjacent to a large-scale artificial lake (30 hectares), with the public green space (50 hectares) and a 27-hole golf course (121 hectares) in surrounding areas. The environment is elegant and traffic is convenient; together with nearby comprehensive projects which are being developed by Poly, Sino-ocean, Wantong and FORTE, it will bring a modern and high-quality environment and draw the curtain for the unlimited potential CBD development.

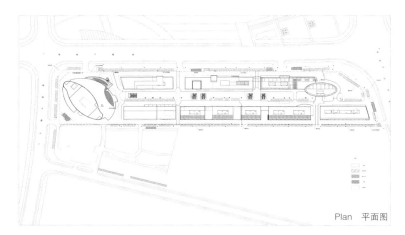

Plan 平面图

Overall design idea—planning concept

HOPSCA planning concept is adopted to combine functions of residence, business office, shopping, entertainment, social intercourse and travel & rest. Ronghe Plaza is a compound architectural complex that is uniform in its overall height and highly compacts with various international elements; it has a total construction area of more than 400,000 m², and its design style is modern, simple, international and fashionable. The main body is composed of six 5A business office buildings, with a construction area of more than 260,000 m²; besides, the project also includes a five-star hotel, luxury apartment, serviced apartment and large-scale underground scene commercial city.

The project combines the ecological and elegant environment of Binhai New Area with the modern and distinctive architectural images, which makes the architecture respond to the environment and realizes positive cycle in the region. Its design fully considers the visual experience and spatial scale of people in the indoor and outdoor public areas, creating a pleasant environment; the design pays attention to the main corners and joints which are diverse but uniform in the form. This project becomes the model of high starting point and high completion degree in Binhai New Area.

Architectural concept

As a new urban combination, Ronghe Plaza possesses a super-excellent geographic position, transverse base plane, 690 m-long main facade along the street facing the central lake and artery, and 2 important corners.

Based on this, two rows of strips with different heights+office building form relatively placid sequence, which is the main characteristic of the planning; the north side is lower, with 2-5 floors and 2 openings, and the south side is higher, with 7-11 floors and 3 openings; they are divided into independent buildings.

At the east and west ends of the project, the curve masses with elliptic shape and mango shape are set as the visual focus, which has caused unique corner building effect and established a romance and elegant landmark in Binhai New Area. The first floor of elliptic building is the public transport junction; building with mango shape is the five-star hotel; they have provided a clear background landscape for the inner street formed by the two strip-shaped complexes.

While pool, garden, glass foot bridge and sightseeing elevator set in the inner street have broken the monotony caused by long block, and each part presents a different world that is full of interest.

The five-star hotel in the west has 11 floors, the typical floor plan is mango-shaped, and its annexes has flexible curved surfaces; the entrance porch full of expressive force turns into a symbolic image at the street corner, and the curve structure responds to elliptic shape at the lower level in the east. Guest rooms are set above the 3rd floor; the 1st and 2nd floor contain public rooms, including conference room, dining room, banquet hall, reception lobby, etc.; the entire form is stretched and transparent seen from the entrance.

Exterior image of the building: modern European style is adopted for southern and northern buildings, modern techniques like mass incision, cantilever and hollowing are employed, and metal plates and glass are used to highlight the hale and hearty atmosphere; various rhythms of vertical windowpanes have virtual-real comparisons, in full glasses or dead walls, which have shown a sculptural beauty and modern sense.

The long facade is decorated with glazed glass in the middle, the dynamic sense and vivid tone can prevent aesthetic fatigue; the curtain wall in dot mode and canopy decorated with perforated panel also add fresh contents to close-range observation.

Such a long complex is just like a complicated symphony in the city, demonstrating rich, elegant and fashionable atmosphere to visitors. Although the building is composed of lifeless materials, it is vital and expresses its life with the composition elements and relations among them, to sympathize with our soul.

"Leave some leeway for possibility", "choose possibility that can be realized"; designers have left all preconceptions aside at the beginning and in the middle of the project, they have taken every part into account with open heart and durable patience (including clients, the government and society), accepted and coordinated all kinds of contradiction and conflict, and turned every possibility into reality through free expression.

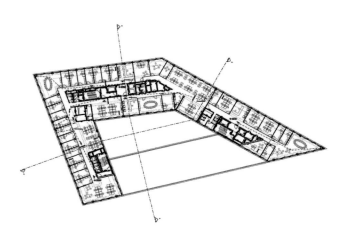

Ericusspitze
Ericus-Contor, 3rd Floor, Scale: 1:350
Henning Larsen Architects A/S

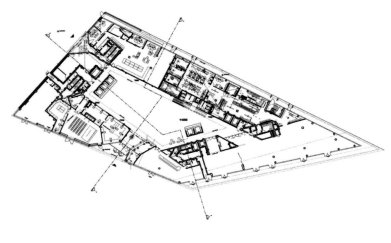

Ericusspitze
Spiegel, Groundfloor, Scale: 1:350
Henning Larsen Architects A/S

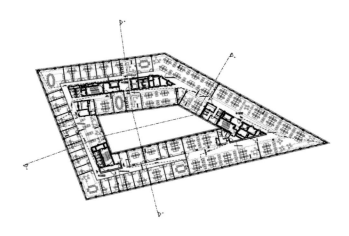

Ericusspitze
Ericus-Contor, 9th Floor, Scale: 1:350
Henning Larsen Architects A/S

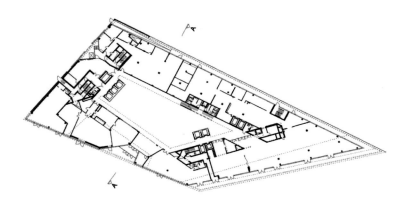

Ericusspitze-Spiegel
Level 0
Scale 1:500

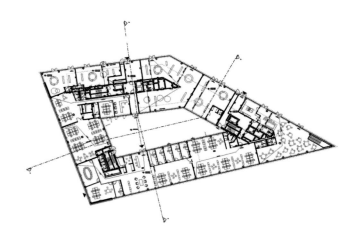

Ericusspitze
Ericus-Contor, Groundfloor, Scale: 1:350
Henning Larsen Architects A/S

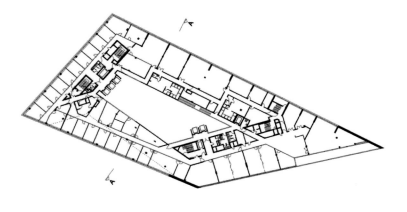

Ericusspitze-Spiegel
Level 1
M 1:500

Plan 平面图

面对 30 多万平方米的建筑体量、600 余米长的街区景观，设计师们在这个拥有不容忽视的巨大影响力和具有积极持续作用的新都市片段中感到强烈的社会责任感以及使命感。

设计师以开放的心态、现代的材料和手法精心打造作品，雕刻着种种主题，演绎着种种细节，试图让新材料、新形式大量积累与爆发，给人精彩纷呈的视觉体验，浓缩出新的地域和时代精神。

这 600 米长的现代建筑诗篇，在日积月累的精雕细刻中，在长达 5 年的时断时续的更改后终于呈现。它在 2011 年空港最受欢迎的十大建筑评比中名列首位。

项目背景

作为滨海新区连接京津两地的西大门，天津空港经济区经过多年的发展，已经形成了航空、电信、装备制造、软件服务外包、总部经济五大产业齐头并进的局面。未来该处将更突出功能复合、生态宜居等特点，形成具有空港特色的高端商圈及高科技的发展氛围，成为"东方西雅图"。

融和广场坐落于天津空港经济区主入口位置的核心地段，与占地 30 公顷的大型人工湖为邻；周边规划有占地 50 公顷的公共绿地、121 公顷的 27 洞高尔夫球场，环境优雅、交通便捷，与附近的保利、远洋、万通、复地等正在开发的综合体项目共同体现现代化、高品质的环境，为潜力无限的空港 CBD 的发展拉开了序幕。

整体构思——规划理念

设计师运用 HOPSCA 规划理念，将居住、商务办公、购物、文化娱乐、社交、游憩等各类功能有机结合。融和广场是一个整体高度统一、国际化元素高度凝练的复合型建筑群，总建筑面积 40 多万平方米，设计风格现代、简约、国际、时尚。主体为 5A 级商务写字楼，规划有六栋，建筑面积 26 万余平方米；此外，项目还包括白金五星级酒店、豪华公寓、酒店式公寓及大型地下体验式情景商业城。

项目将生态、优雅的新区环境和现代化的、特色鲜明的建筑形象巧妙结合，形成建筑与环境的互动与对话，实现区域的良性循环。设计充分考虑人在室内和室外公共活动空间的视线直观感受和空间尺度，营造宜人的环境；注重主要转角节点的塑造，形式多样而统一，成为新区高起点、高完成度的工程典范。

建筑构思

作为新城市结合体，融和广场享有绝佳的地理位置，横向的基地平面使其具有面向中心湖和主干道的 690 米长的沿街立面及两个重要的转角位置。

基于此点，规划的主要特色就是两排高度不同的长条形和办公建筑形成相对平和的序列；北侧为较低部分，高度为 2~5 层，设置了两个开口；南侧为较高部分，高度为 7~11 层，设置了三个开口，分成各自相对独立的楼座。

项目的东西两端分别以椭圆形象和芒果形平面的曲面体量创造了视觉焦点，带来独特的街角建筑效果，为新区树立具有可识别性的浪漫、含蓄典雅的地标建筑。椭圆平面建筑一层对应公共交通枢纽，芒果形平面的建筑则对应五星级酒店；这为两个条形综合体形成的内街提供了风格清新的底景。

而内街中设置的水池、花圃、玻璃天桥、观光电梯等则打破了长向街区所易导致的单调，使每个段落都情趣盎然、各有天地。

西端五星级酒店高度为 11 层，标准层平面为芒果状；裙房则为自由的曲面，配合富有表现力的硕大入口门廊，在街角产生标志性形象；曲线造型与东端低层部分的椭圆形体相呼应。酒店三层以上为客房部分，一、二层为公共用房，有会议室、餐厅、宴会厅、接待大堂等服务用房，从入口看去，具有舒展而通透的整体形态。

建筑的外部形象：南北两排板式建筑采用现代欧式风格，以自由的体量切割、悬挑、挖空等现代手法，配合金属板和玻璃等材料，突出硬朗气质；竖向窗格呈现各种韵律，局部有大的虚实对比，一般为全玻璃或全实墙，为建筑增强玻璃雕塑感、现代感。

长向立面中间点缀彩釉玻璃，带来的跳动感和鲜艳色调可以避免审美疲劳；局部点式幕墙、打孔板装饰的雨棚等使近距离观摩也有新鲜内容。

如此之长的综合体犹如一首复杂的都市交响曲，向前来造访的人们展示着丰富、雅致与时尚的气息。建筑虽然由无生命的材料建成，但通过各种元素及它们之间的关系来表达出生机，并与人的心灵产生了强大的共鸣。

"给可能性留有余地"、"挑选可以实现的可能"，设计师在项目开始和进行之时，抛开一切先入之见，以开阔的心胸、持久的耐性，最广泛地顾及各个方面的需求（包括客户、政府、社会），接受与协调各种矛盾与冲突，利用表达的自由，将可能性转化为现实。

SHANGHAI SHIMAO INTERCONTINENTAL "WONDERLAND" HOTEL LANDSCAPE DESIGN REVIEW

上海世茂新体验洲际酒店景观方案设计回顾

Architecs: Atkins
Location: Shanghai, China
Area: 58,494 m²

设计机构：Atkins 公司
项目地点：中国上海市
面积：58 494 平方米

Project information

Shanghai Shimao Intercontinental "Wonderland" Hotel is located in the western suburb of Shanghai City—Songjiang District. The site has convenient transportation links as it is approximately 40 km from the center of Shanghai, 18 km from Hongqiao Airport, and 75 km from Pudong International Airport. It neighbors the Hu-Qing-Ping Expressway, the Shanghai—Hangzhou Expressway, the A30 high-speed rail line and subway line 9.

Songjiang is the birthplace of Shanghai's history and culture. It is picturesque and historically known for the "Jiufeng sanmao". "Jiufeng" means nine peaks. Ranked as the top of the nine peaks, the Tianma Mount and Heng Mount are located in the north and west of the site. While the Hengshan Pond and Wangjia Pond also flow through the north and west of the site. The rocks, cliffs and water within the quarrying pits even reflect the characteristics of the "Jiufeng sanmao"—nine peaks and three ponds.

The quarrying pit is approximately 100 m deep, 240 m long and 160 m wide. It forms an inner lake within the pit. The pit is surrounded by grass paddy fields and idyllic landscape. A five-star hotel will be built in the pit in the future.

Design concept

Through analysis of the site, the soul of the architectural design is extracted. The "forest", "mist", "water" and "stone" are the souls. The architectures do not intend to destroy the quiet nature of the site, but set the starting point for the design concept; to create a hotel which blends naturally into the surrounding nature.

The open natural environment, spectacular quarries, and beautiful mountains have left a deep impression on all visitors. In order to sustain these impressive and awe-inspiring elements, the hotel is designed to be located inside the pit. Only a few buildings will remain on ground level to fulfill the demand for hotel transport and public functions.The hotel's main body in the pit accommodates hotel rooms, dining rooms above and under the water surface, and a fitness and leisure centre. It extends along the cliff surface, combining with flexible elements to ensure rooms natural growth and evolution of the development. The landscape design needs to coordinate with the architectural style and provides different functional requirements for the development.

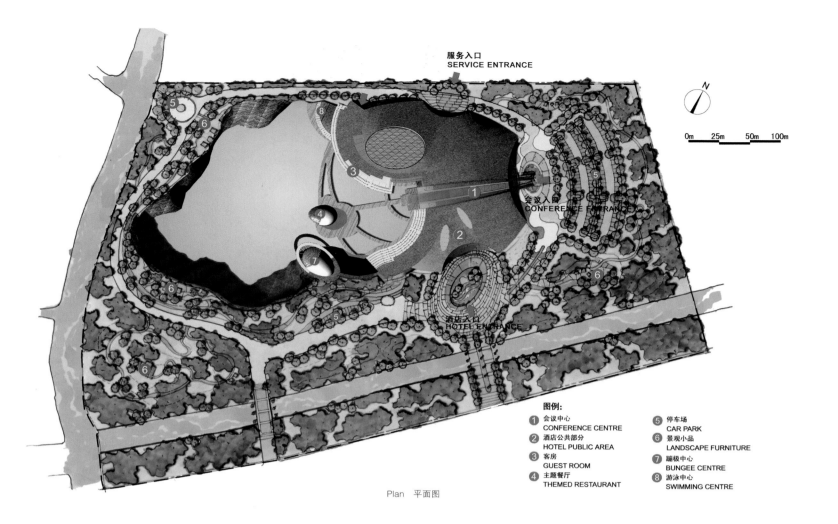

图例:
① 会议中心 CONFERENCE CENTRE
② 酒店公共部分 HOTEL PUBLIC AREA
③ 客房 GUEST ROOM
④ 主题餐厅 THEMED RESTAURANT
⑤ 停车场 CAR PARK
⑥ 景观小品 LANDSCAPE FURNITURE
⑦ 蹦极中心 BUNGEE CENTRE
⑧ 游泳中心 SWIMMING CENTRE

服务入口 SERVICE ENTRANCE

会议入口 CONFERENCE ENTRANCE

酒店入口 HOTEL ENTRANCE

Plan 平面图

Landscape design

Landscape design plays an important role in this quarry hotel design. The theme of the entire landscape design is about "nature".

The landscape of the hotel is divided into two parts—the ground level and the bottom of the pit. The landscape design of the ground level mainly uses planting to restore a natural landscape environment. A main walkway is designed to surround the quarry. Rivers and buildings create many interesting nodes, shaping the natural landscape of the area.

The landscape design at the bottom of the pit is composed of water features and vertical greening. The different layers of falling balcony gardens were combined with the main hotel building creating a unique ecological environment that penetrates to the inside of the building, creating a link between the building and the landscape. The artificial vertical greening of the cliff and the waterfall will create a spectacular vertical screen for the hotel guests. During the night, lighting effects will project onto the cliff and increase the mystery of the pit. Since there is natural groundwater at the bottom of the pit and an artificial outdoor swimming pool, landscape designers have tailored two different water gardens—"outdoor hidden water garden" and "mist water garden".

The outdoor hidden water garden is filled with tropical style planting. The aim is to create a hotel garden located in a natural valley. There are trimmed box planting along the artificial pontoon, and bars and some casual chairs are provided for residents to relax and enjoy their surroundings.

The natural water body within the pit will become a mist water garden. Small streams and brooks will flow through the rocky landscape, creating a natural and peaceful space. Natural aquatic plants are planted in shallow water in order to improve the visual quality of the pit, creating a more intimate effect for people in the atrium. Meanwhile, it also provides an independent growth environment for the aquatic plants while also realizing water purification.

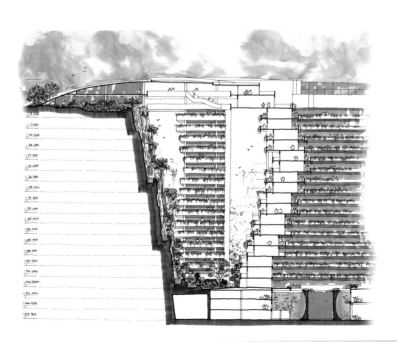

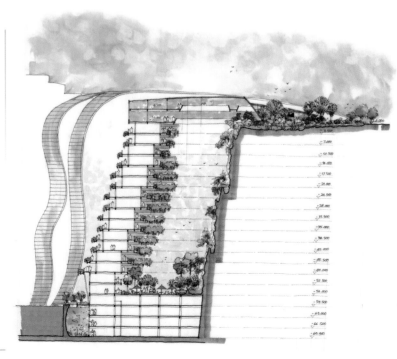

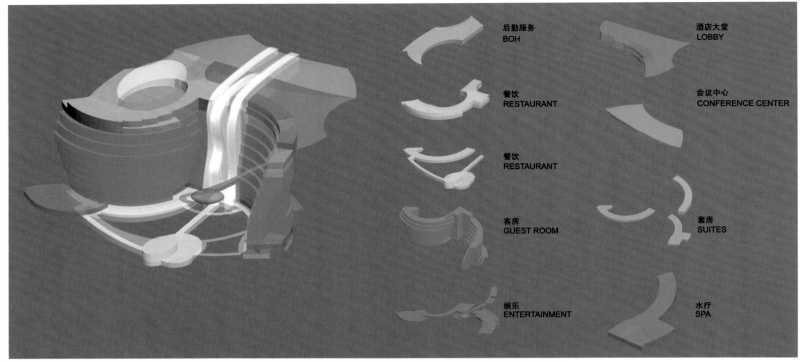

后勤服务 BOH
餐饮 RESTAURANT
餐饮 RESTAURANT
客房 GUEST ROOM
娱乐 ENTERTAINMENT

酒店大堂 LOBBY
会议中心 CONFERENCE CENTER
套房 SUITES
水疗 SPA

Analysis　分析图

项目概况

上海世茂新体验洲际酒店位于上海西郊古城——松江。基地交通便捷，距离上海市中心约40公里，虹桥机场约18公里，浦东国际机场75公里。周边邻近沪青平高速、沪杭高速、A30高速及轨道交通9号线。

松江，是上海历史文化的发祥地，山明水秀，历史上素以"九峰三泖"著称，位列九峰之首的天马山和横山分别位于基地的北部和西部；而横山塘、旺家浜亦流经基地的北面和西面，基地采石深坑内的岩石、崖壁和水更是突出了"九峰三泖"的特色。

这个采石坑深约100米，长240米，宽160米左右，形成内湖。深坑周围为草地水田，一派田园风光，未来的酒店就建造于这个深坑之中。

设计构思

通过对基地的分析，为建筑方案提炼出设计的灵魂——"林"、"蕴"、"水"、"石"。建筑师们无意破坏宁静的自然，为设计构思定下了出发点——"创造一个源于自然而又融于自然的绿色度假酒店"。

自然环境的开阔、采石坑的壮观和山峰的秀美，给所有的到访者留下了深刻的印象。为了能将这些令人感动和震惊的元素延续和保留，设计中将建筑主体布置于深坑中，在地面上仅保留少量必需的建筑，以满足酒店的交通和公共功能的需求。坑内的建筑容纳了酒店客房、水面和水下特色餐饮、健身与休闲中心等功能。建筑依靠岩壁以最大面展开，结合柔性要素，寻求自然的生长和演变。景观设计需要与建筑风貌相配合，同时依照功能区的不同特征，结合周边环境，合理安排景观元素。

景观设计

景观设计在本案的设计构思的体现上起到了重要作用。整个景观设计的主题都围绕"自然"主题开展。

酒店的景观主要分为地面部分和深坑底部两大部分。地面部分的景观设计以软景绿化为主，以还原一个自然的地貌环境。基地内部步行景观线路采用自然生长的方式围绕着采石坑。河流和建筑创造出多个趣味节点，塑造自然而丰富的区内景观。

深坑底部的景观以水景和垂直绿化为主。室外层层跌落的空中花园与建筑主体相结合，为室内创造了独一无二的生态环境，做到室内外的渗透和贯穿。崖壁的人工绿化以及跌落的瀑布，为酒店的住客打造了一个壮观的垂直画面。同时夜晚映射在崖壁上的灯光效果，更是增加了深坑的神秘感。由于深坑下面有一个自然形成的地下水内湖，结合酒店的室外人工泳池，景观设计师们量身定做了两种不同的水景花园——室外隐水花园和云雾水景花园。

室外隐水花园充满了热带花园的风格，它结合泳池的水面景观，意图创造一个坐落在自然山谷里的酒店花园。两侧人工种植的植物箱沿着人工浮桥布置。这里有吧台以及一些休闲座椅，供住客们观赏和休息使用。

对于深坑内的自然水体部分，景观设计师们设计了云雾水景花园，旨在创造一种薄雾以及小水流流过岩石的景观。在窄小的水面上配备自然的水生植物，以扩大整个深坑水面的延伸感，为整个中庭创造一个较为私密的空间。同时为水生植物在此间的生长提供了一个独立的生长环境，以达到水体净化的效果。

NARADA RESORT & SPA PERFUME BAY
海南香水湾君澜度假酒店

Architecs: Allied Architects International
Cooperative design: Zhejiang Prov. Institute of Architectural Design and Research
PAL Design Consultants Ltd.
Edfa Design Limited
Client: Zhejiang Narada Hotel Management Co., Ltd.
Location: Perfume Bay, Hainan, China
Base area: 315,000 m²
Total construction area: 29,270 m²
Photographer: Jin Zhan, Kerun Ip

设计机构：AAI 国际建筑师事务所
合作设计：浙江省建筑设计研究院
          PAL 设计事务所有限公司
          易道国际规划设计有限公司
客户：浙江世贸君澜酒店管理有限公司
项目地点：中国海南省香水湾
基地面积：315 000 平方米
总面积：29 270 平方米
摄影：金露、Kerun Ip

Narada Hotel has an excellent natural environment, with its design based on the sea. Tourists will start their unforgettable vacation at the moment when they step into the hotel lobby. Therefore, visual experience directly associated with the sea leaps to the eyes at the entrance of the hotel.

This is a villa-style resort hotel with more than 100 guest rooms, and meantime it holds three dining rooms, a bar, a hydrotherapy center, a small conference center, a fitness center, and a unique double-deck open-air swimming pool.

A 150 m long waterscape axis is lined out by AAI on the site, in which a swimming pool with a length of 88 m and a width of 8 m directly extends to the sandy beach from the reception area and it seems to be linked together with the horizon viewed from the water surface. All public service facilities are arranged along both sides of the principal axis, covering conference center, hydrotherapy center, dining room, etc. It not only constitutes a favorable spatial organization for public elements, but also creates an independent architectural complex with distinct characters which has covert garden or open seascape.

A wooden door becomes the doorway connecting public areas with private areas. People can reach the super-scale guest room where they can enjoy the half-open bathroom with a large block of glass by going through the garden path, walking around the relaxation pavilion, and passing the swimming pool. Every journey will give you a different experience. Different villa types, independent dwellings, courtyard dwellings or combined dwellings, provide more options and different feelings for visitors. At the same time, in order to provide beautiful seascape for villas far away from the seaside, architects have piled up three platforms with different elevations on the flat grounds before. Villa guest rooms are scattered within it, partly visible among the intense tropical trees.

Architects have deconstructed the representative residences in China, adopted some elements to form a construction menu on the premise of maintaining environmental scale, such as sloping roof, white wall, rectangular opening and pillar, and then reorganized them in a modern way. In addition, they have also analyzed some unique details such as the roof, and added some special elements.

North Elevation　中轴线北立面图　　0 5 10 15 20 25M

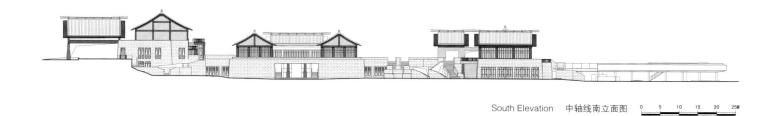

South Elevation　中轴线南立面图　　0 5 10 15 20 25M

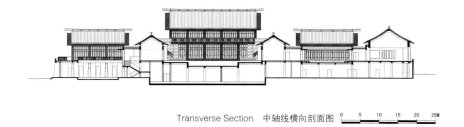

Transverse Section　中轴线横向剖面图　　0 5 10 15 20 25M

Longitudinal Section　中轴线纵向剖面图　　0 5 10 15 20 25M

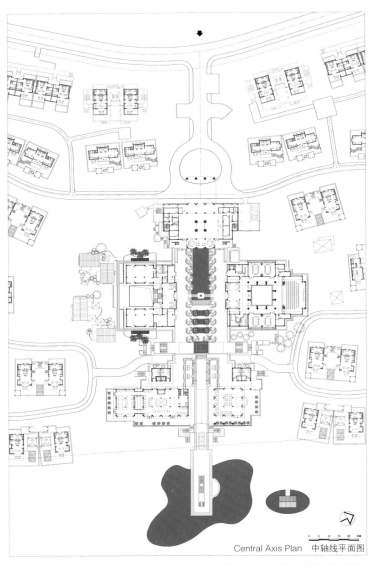

Central Axis Plan 中轴线平面图

君澜酒店有着优越的自然环境条件，以大海本身作为设计根本，让游客从进入酒店大堂的那一刻，就开启了一次难以忘怀的假期经历。因此，在酒店的入口处，映入眼帘的就是与大海直接相通的视觉体验。

它属于别墅式度假酒店，拥有百余间客房，同时容纳了三个餐厅、一个酒吧、水疗中心、一个小型会议中心、一个健身中心以及一个独一无二的双层室外游泳池。

AAI 在场地上划出一条近 150 米长的水景中轴线，其中有一个 88 米长、8 米宽的泳池，从接待区域直接延伸至沙滩，从水面望去，仿若海天相连。沿着主轴的两边布置了所有的公共服务设施，包括会议中心、水疗中心、餐厅等，不仅为公共元素构成了良好的空间组织，也创造出个性鲜明的独立小型建筑群，使其拥有隐秘的小庭院或开阔的海景。

一道木门成为连接公共区和客房私密区的出入口——人们经由花园小径、转过休息亭、绕过泳池、进入超尺度的客房、享用大玻璃半开放式的卫生间，每一次停转都会有不一样的体验。独院、合院或不同的别墅类型让访客有更多的选择及不同的感触。同时，为了使远离海边的别墅客房都能拥有美丽的海景，建筑师在原本平坦的场地上堆出了三个不同标高的台地，别墅客房就散落其间，在浓郁的热带树林掩映下若隐若现。

建筑师对中国具有代表性的住宅进行解构，在保持环境尺度的同时，撷取部分元素组成一个建筑菜单，如坡屋顶、白墙、长方形开口以及廊柱，并以现代的方式进行重组。此外，还对一些独特的细节比如屋顶进行分析并加入特别的元素发展，创造出既有地域性又清雅脱俗的建筑效果。

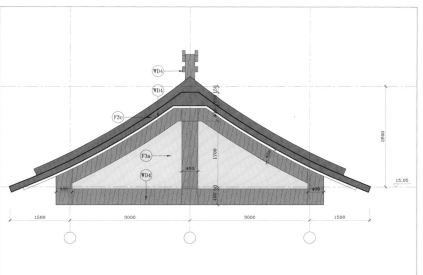

Analysis chart 1　分析图 1

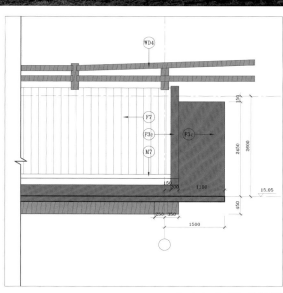

Analysis chart 2　分析图 2

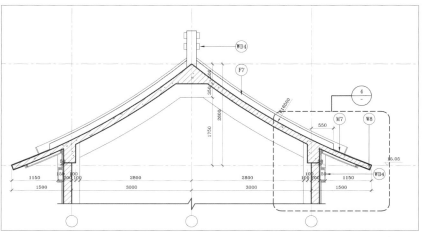

Analysis chart 3　分析图 3

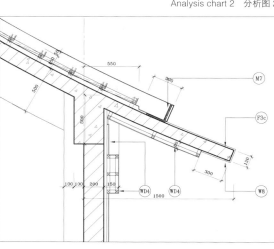

Analysis chart 4　分析图 4

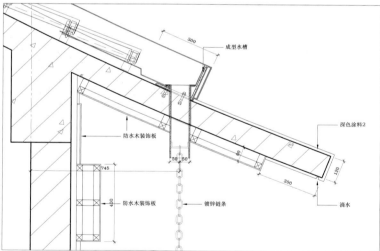

成型水槽

深色涂料2

防水木装饰板

防水木装饰板

镀锌链条

滴水

Analysis chart 5　分析图 5

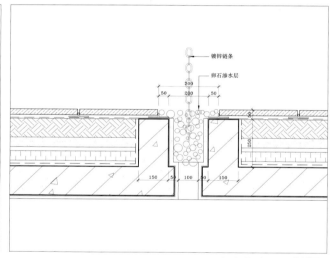

镀锌链条

卵石渗水层

Analysis chart 6　分析图 6

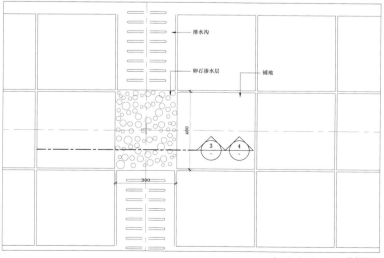

排水沟

卵石渗水层

铺地

Analysis chart 7　分析图 7

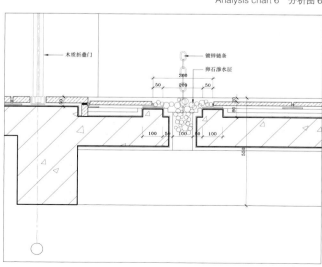

木质折叠门

镀锌链条

卵石渗水层

Analysis chart 8　分析图 8

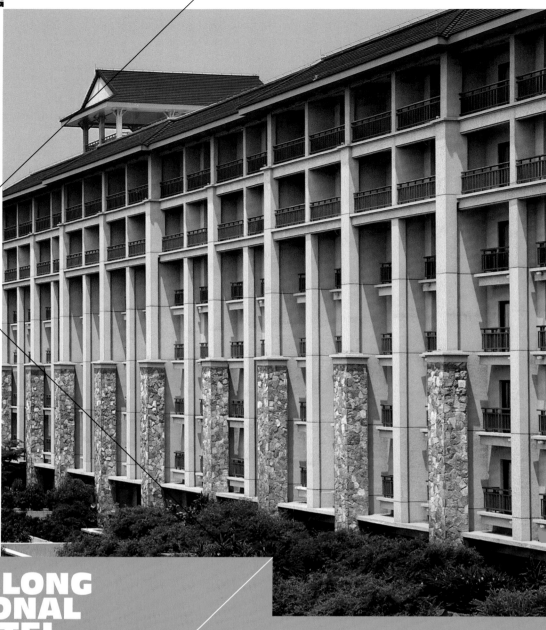

NEW CHIMELONG INTERNATIONAL RESORT HOTEL

新长隆国际度假酒店

Architects: Beijing Sino-Sun Architectural Design Co., Ltd.
Cooperative design: NEWSDAYS
Location: Guangzhou, China
Total area: 168,808 m²
Photographer: Nigel Young

设计机构：北京东方华太建筑设计工程有限责任公司
合作设计单位：广州集美组室内设计工程有限公司
项目地点：中国广州市
总面积：168 808 平方米
摄影：Nigel Young

The five-star New ChimeLong International Resort Hotel is located in the center of Guangzhou ChimeLong Group and bands together with the original ChimeLong Hotel. It has a building site area of 140,115 m² and a total construction area of 168,808 m²; the new hotel has 1,068 guest rooms, an international exhibition conference center and supporting service facilities which can provide services for customers.

The design purpose of this project is to create a new holiday life. The unique theme is emphasized through four concepts which are jungle, animal, castle and Lingnan. The position superiority is fully utilized in this design: most guests enter the hotel by passing through the main entrance at the northeast corner of the big castle surrounded by international convention center, group guest room building, lobby and animal island and the original hotel, to start their mysterious journey. Around this entrance, vegetative landscape and mountain slope of intense tropical rain forest have formed a buffer zone between the building and the road. The hotel lobby can be reached by passing the avenue decorated by animal sculptures from the main entrance, and the square surrounded by torches will be the first impression left on visitors when they enter the castle. After entering the lobby space through a wooden bridge, your holiday life will start. The wonderful mood during holiday will be enhanced by the white tiger garden. The lobby space in the hotel has a height of 20.5 m; sunlight pours down through the skylight, which will give more pleasure to you.

The style of guest room is wild but elegant, containing Hakka spirit of Lingnan and reproducing the ancient style. This is a new perspective of the design and also creates the new style of ChimeLong. Sloping roof and rhythmic guest room buildings in ladder type comply with the original mountain shape. Lobby and conference center in modern castle style of Lingnan provide distinguished experience and a dream world for the guests.

番禺长隆度假酒店扩建

The whole architectural complex is bathed in the genial sunshine, surrounded by the green tropical rainforest, and accompanied with birds chirping and breeze. As it were, the whole project itself is a fictitious land of peace. It completely cuts off the noise and pressure in cities and creates a new vacation pattern. New ChimeLong International Resort will be a unique, wide, noble and private world, and is destined to be a first-class theme resort. The unique environment and luxury facilities will provide an unforgettable experience for all visitors.

五星级新长隆国际度假酒店地处广州新名片长隆集团的中心地段，与原长隆酒店紧密结合在一起。建筑用地140 115 平方米，总建筑面积 168 808 平方米。新酒店共有 1 068 间客房和国际会展会议中心及配套服务设施等，可以为客户提供良好的服务。

此项目的设计意图是创造一种全新的度假生活。通过丛林、动物、城堡、岭南这四个概念强调项目的独特主题。设计充分利用场地及其方位优势，大部分客人通过由国际会议中心、团体客房楼、大堂及动物岛客房楼、原有酒店所围合成的大城堡的东北角主入口进入酒店，开始他们激动人心的神秘旅程。在这个主入口周围，浓郁的热带雨林植物景观和山坡构成建筑和道路之间的缓冲带。可以从主入口经由动物雕塑点缀的星光大道到达酒店大堂前，火炬环绕的广场带给游客进入城堡的第一印象。经过木桥进入大堂前厅空间，由此拉开度假生活的序幕。度假的美妙心情也因见到白虎庭院而有所增强。酒店大堂中心空间高达 20.5 米，阳光从顶上的天窗倾泄而下，使得度假的心情更加愉悦。

客房的风格狂野而高贵，蕴涵岭南的客家围楼精神，重现远古的呼唤。这是设计的一个新视角，也再次创造了长隆新风格。坡屋顶和有节奏的阶梯状客房建筑顺应了原有的山势。现代岭南城堡风格的大堂及会议中心让尊贵的客人体验到了从未有过的梦幻感觉。

整个建筑群沐浴在和煦的阳光中，被绿色的热带雨林环绕，伴随着小鸟的鸣叫，清风微拂。可以这么说，整个项目本身就是一个世外桃源，它把城市的喧闹与压力完全隔绝，为游客创造了一种新型体验式度假方式。新长隆国际度假酒店将是一个独特、狂野、高贵且私密的世界，也将是一个一流的主题度假村。独特的环境和豪华的设施将为所有到访者提供难以忘怀的体验。

Plan　平面图

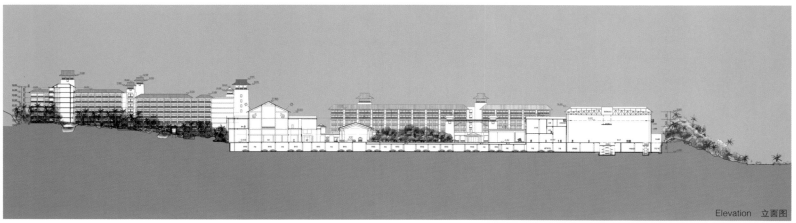

Elevation 立面图

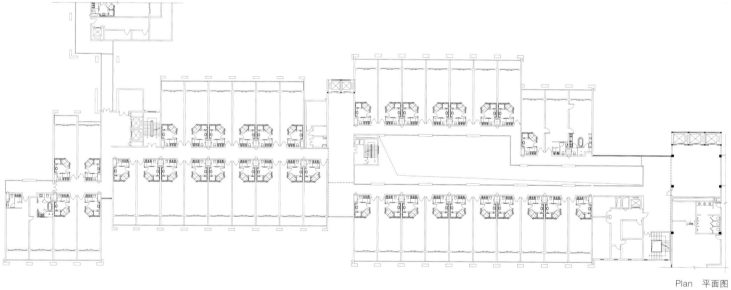

Plan 平面图

GOLDCREST TOWER
DUBAI, UAE
阿联酋迪拜 Goldcrest 大厦

Architects: Atkins
Location: Dubai UAE
Area: 57,231 m²

设计机构：Atkins 公司
项目地点：阿联酋迪拜
面积：57 231 平方米

Goldcrest Tower is prominently located on the waterfront of the prestigious Jumeirah Lake Development in Dubai. Atkins was responsible for architectural, MEP and structural design services as well as construction supervision. This project was commissioned not long after the design of a sister Goldcrest residential tower. The design comprises a 41-storey tower above a podium deck and three basement below ground. The dramatic visual and spatial forms adopted have combined aesthetic composition to provide a contemporary style embodied in the simple balanced proportions of the building. Externally, the strength of the design concept emanates from two tapering fin elements which curve in two directions. The fins form the boundary of the building at the lower level and provide support and strength for the residential blocks. A careful incorporation of colour, shadow, texture and materials has been employed to provide a rich balance of composition which is enhanced by the clean horizontal lines of the residential balconies. Internally, office space is designed around a rectangular core, offering maximum views to the surrounding lakeside. The central core is designed to accommodate the lifts, staircases and service areas. Two levels are plant rooms. Both the residential and the commercial tower have been completed.

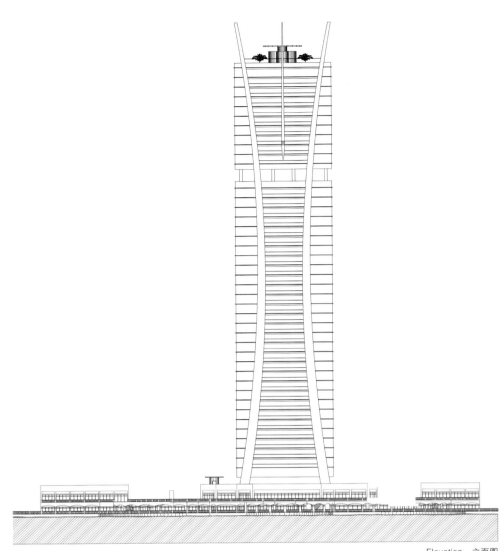

Goldcrest 大厦坐落在迪拜朱美拉湖湖滨开发区的显著位置，Atkins 负责本项目的建筑设计、工程和结构设计以及工程的监理工作。本项目是在 Goldcrest 住宅楼设计之后不久进行设计的，项目由裙房上部 41 层的塔楼以及 3 层地下室组成。引人注目的视觉和空间形式结合了超凡美感，现代的建筑风格简约地平衡了建筑各部分的比例。在外形上，建筑设计的力量来源于沿两个方向弯曲的逐步变细的鱼鳍状元素。鳍结构在低楼层形成了建筑的外部边界，并且为相邻的住宅板块提供支撑。悉心搭配的色彩、阴影、结构和材料使建筑组成丰富又相互平衡，使呈水平线条的住宅阳台更加突出。在建筑内部，办公空间设计在矩形的核心筒周边，以最大限度地展露周边的湖景。核心筒结构用来提供电梯、楼梯和服务空间。建筑有两层专门用来培养植被的房间。

目前住宅塔楼和商业塔楼均已完工。

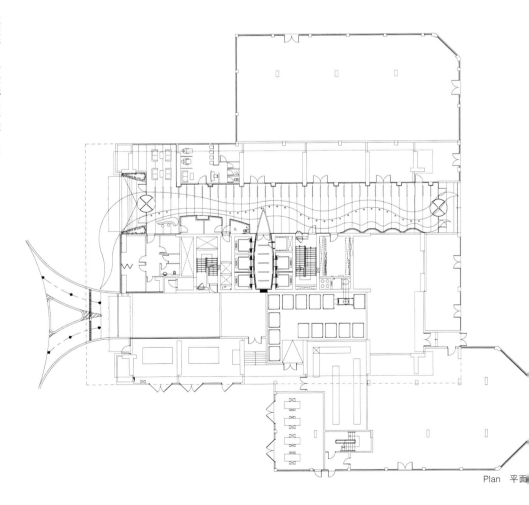

MILLENNIUM TOWER
DUBAI, UAE
阿联酋迪拜 Millennium 大厦

Architects: Atkins
Location: Dubai, UAE
Area: 99,800 m²

设计机构：Atkins 公司
项目地点：阿联酋迪拜
面积：99 800 平方米

Atkins was appointed as chief consultant for architectural, interior, structural, mechanical and electrical engineering design, construction supervision and project management for this 285 m-high, 59-storey tower on Sheikh Zayed Road comprising two- and three-bedroom apartments, a multi-storey car park, health facilities and retail outlets. The Millennium Tower is located on a prominent plot adjacent to the Sheikh Zayed Road in Dubai and rises to 285 m. The client's dual requirements were aesthetics and functionality. The client's brief demanded a modern building which was reminiscent of the 20th century art deco movement that resulted in the design of such icons as the Chrysler Building and the Empire State Building. This tower incorporates what looks like a sliding mass that is essentially achieved by slipping different elements over one another in vertical planes in a tapering fashion so that the lowest mass rises to the topmost of the building. A solid frame element anchors the building to the ground whilst a second central body supported by the frame element expresses a vertical upward movement. The articulation of the main body into lighter elements at the apex of the building is further emphasised by the presence of the deep recess at the base which creates a sense of weightlessness to the "main body" of the tower. The body glass, which is non-reflective, is tinted with fifty percent transparency to afford some view of the interior thereby creating a sense of animation to the facade. The apartment layouts are standardised to maximise functionality and cost effectiveness. A total of 301 three-bedroom apartments and 106 two-bedroom apartments are provided on 55 typical floors, and the building services are located on floors 10, 30 and 50. Parking facilities, for 471 vehicles, are provided in the multi-storey car park building, the roof of which accommodates a 25 m-long swimming pool, a gymnasium, squash courts and changing rooms.

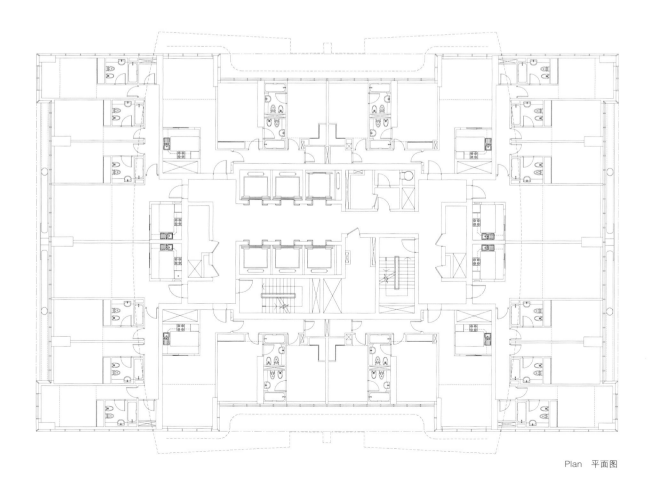

Plan 平面图

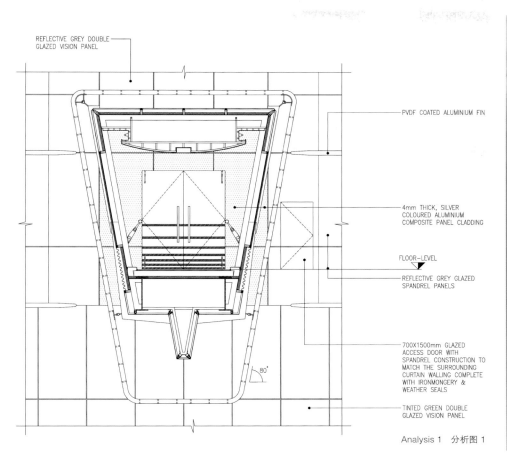

REFLECTIVE GREY DOUBLE GLAZED VISION PANEL

PVDF COATED ALUMINIUM FIN

4mm THICK, SILVER COLOURED ALUMINIUM COMPOSITE PANEL CLADDING

FLOOR-LEVEL

REFLECTIVE GREY GLAZED SPANDREL PANELS

700X1500mm GLAZED ACCESS DOOR WITH SPANDREL CONSTRUCTION TO MATCH THE SURROUNDING CURTAIN WALLING COMPLETE WITH IRONMONGERY & WEATHER SEALS

TINTED GREEN DOUBLE GLAZED VISION PANEL

80°

Analysis 1　分析图 1

　　Atkins 公司被指定作为迪拜 Millennium 大厦的建筑设计、室内设计、结构设计、机械和电气工程设计、工程监理以及项目管理的首席顾问。位于著名的 Sheikh Zayed 大道上的这座 285 米高、59 层的建筑包括两居室和三居室公寓、多层停车场、保健设施和零售商店。迪拜 Millennium 大厦毗邻 Sheikh Zayed 大道的著名景点，使大道的天际线上升至 285 米。客户对于美学和功能有双重要求，他们希望看到一个能让人想起在 20 世纪因装饰艺术运动而建设的克莱斯勒大厦和帝国大厦的现代建筑。整个塔楼看起来像一个滑块，上部逐步变细的形式使其在垂直方向上看似通过不同的元素彼此滑落而成。一个实体框架将建筑固定在基础上，同时框架支撑的第二个中心体展示了一种垂直向上的运动。当主要建筑主体在顶端体量变轻后，建筑通过厚重的基础在塔楼的主体上形成一种"失重感"。建筑的非折射玻璃有 55% 的透明度，可以令人隐约看到建筑的内部，从而形成建筑外立面的动态感。建筑公寓的布局是标准化的，从而使建筑功能和成本效率最大化。建筑在 55 层中总共提供了 301 间三居室公寓和 106 间两居室公寓，物业服务设在第 10 层、30 层和 50 层。建筑的多层停车场可容纳 471 辆轿车，其顶部有一个 25 米长的游泳池、一个健身房、壁球场及更衣室。

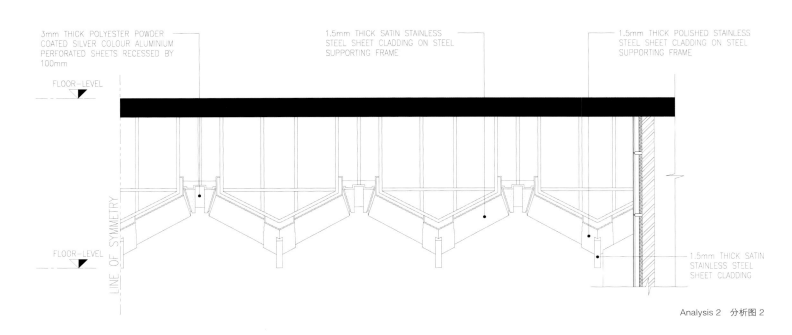

3mm THICK POLYESTER POWDER COATED SILVER COLOUR ALUMINIUM PERFORATED SHEETS RECESSED BY 100mm

1.5mm THICK SATIN STAINLESS STEEL SHEET CLADDING ON STEEL SUPPORTING FRAME

1.5mm THICK POLISHED STAINLESS STEEL SHEET CLADDING ON STEEL SUPPORTING FRAME

FLOOR-LEVEL

FLOOR-LEVEL

LINE OF SYMMETRY

1.5mm THICK SATIN STAINLESS STEEL SHEET CLADDING

Analysis 2　分析图 2

HAINAN WANNING SHIMEIWAN WESTIN FIVE-STAR HOTEL
海南万宁石梅湾威斯汀五星级酒店

Architects: Pan-Pacific Design Group
Location: Haikou, China
Area: 79,035 m²

设计机构：泛太平洋设计集团
项目地点：中国海口市
面积：79 035 平方米

Environmental friendly holiday village is built by following people-oriented concept and exploring sustainable use of resources. When tourists' demands for luxury holiday are met, local social and economic development as well as environmental protection should also be promoted. Therefore, design fundamentals of this five-star hotel project is confirmed: natural scenery is set as the principal line, ecological technology is adopted as the main method and the Southeast Asian style is combined with modern construction, to strive for harmony between human and nature, tourists and local community, as well as hotel architectural style and local architectural style; thus sustainable development of the hotel and area where it is located can be promoted.

遵循以人为本的理念，探寻资源的永续利用，建设环境友好型度假村，除了要满足旅游者的豪华度假需求，亦要促进当地的社会经济发展和环境保护。因此，我们确定了该五星级酒店项目的基本设计原理：以自然风光为主线，以生态技术为主要手段，以东南亚式风格与现代建筑相结合为基础，力求达到人与自然界的和谐、旅游者与当地社区的和谐、酒店建筑风格与当地原有建筑风格的和谐，以促进酒店及其所在地区的可持续性发展。

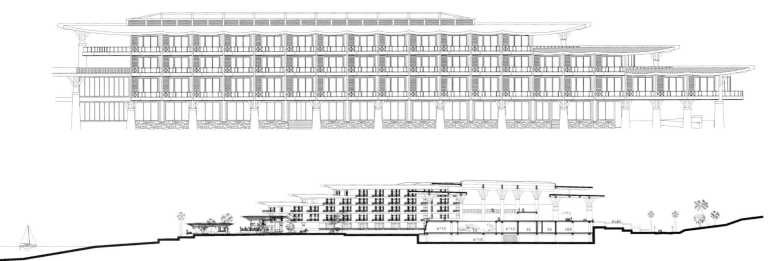

Section 1　剖面图 1

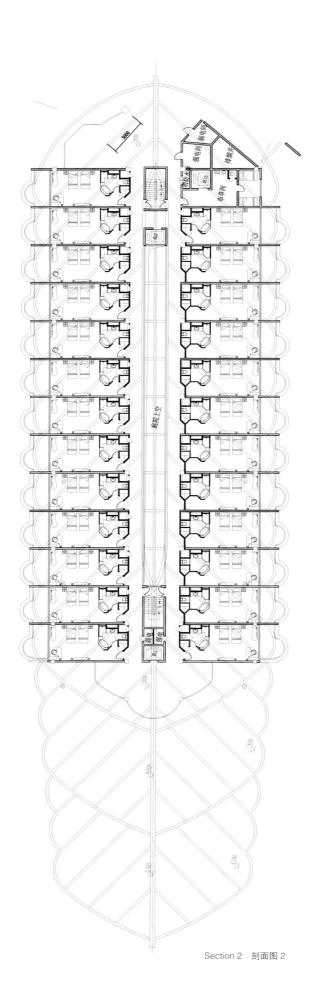

Section 2　剖面图 2

SM NORTH EDSA SKYGARDEN
SM 集团北部空中花园设计方案

Architects: Arquitectonica
Interior designer: Arquitectonica
Landscape designer: Aquitectonica-Geo
Partners-in-Charge of Design: Bernardo Fort-Brescia, FAIA and
Laurinda Spear, FAIA, ASLA, LEED AP
Site area: 26,200 m^2
Gross floor area : 4,510 m^2
Location: Quezon City, Philippines

设计机构：Arquitectonica 建筑事务所
室内设计：Arquitectonica 建筑事务所
景观设计：Aquitectonica-Geo
协助设计： Bernardo Fort-Brescia, FAIA 和
Laurinda Spear, FAIA, ASLA, LEED AP
场地面积：26 200 平方米
总面积：4 510 平方米
项目地点：菲律宾奎松市

Context and aims
The "Skygarden" introduces a green theme to SM Prime's very first mall. The 402 m-long open-air elevated garden will be a heaven where shoppers and visitors can stroll along a landscaped and covered path lined with lush tropical trees, flowering plants and assorted shrubs. The Skygarden stretches from one end of the mall to the other with a number of entrances leading to the block, the city center (main mall) and annex. In effect, this garden will be part of a public circulation loop that connects the outdoor with the indoor. When fully completed, this will be an added attraction to the mall, while reflecting a growing interest in sustainable design around the world.
Planning and massing
Included among the features of the garden are pools & fountains, an amphitheater for shows and special events, F & B pavilions, casual dining spots, tunnel shops and cafes. A waterfall which cascades to ground level promises a colorful scene along EDSA especially when illuminated at night. Additionally, sculptural art pieces, some interactive sketches, are a visual treat at every turn and twist of the covered footpath. Beneath the Skygarden is a public transport depot and additional parking for shoppers.
Round activity areas and dining pavilions are consistent with the circular skylights inside the mall, while the undulating floor plate and footpath are a carry-over from the facade cladding and the ribbon-like playfulness in the annex atrium. All together, they create a dramatic composition which gives the mall an architectural identity not only distincting from SM's other malls, but also being an iconic image that will set it above and apart from the competition.

Landscape strategy

The concept behind the design was to tie the existing malls to an outdoor pedestrian shopping/ dining whose structure, shapes and colors contribute to a total outdoor park experience. Water is used strategically as a vital element to buffer nearby traffic noise and links program elements providing a sensory richness to the project.

The site will be highly planted throughout. The gardens therefore become tied with the restaurant venues on the project and play a central role in creating spaces for the public enjoyment.

The vitality of the site is therefore established via spaces where people and gardens come together and where the green areas and circular plazas/ performances spaces contribute to a total outdoor experience.

Working in conjunction with the pedestrian experience, the covered walk and the tunnels are used to create environments where shopping and dining can be appreciated outdoors, and where spaces range from the more intimate to the very public in a park environment that communicates quality and comfort.

Summary

The design of the Skygarden, together with the total development of SM North EDSA, is a response to the client's requirements for efficiency, functionality, flexibility and the recent focus on sustainable design. Taken together with the block, the city center and annex, the Skygarden becomes an exciting and modern destination for mall visitors in the metropolis.

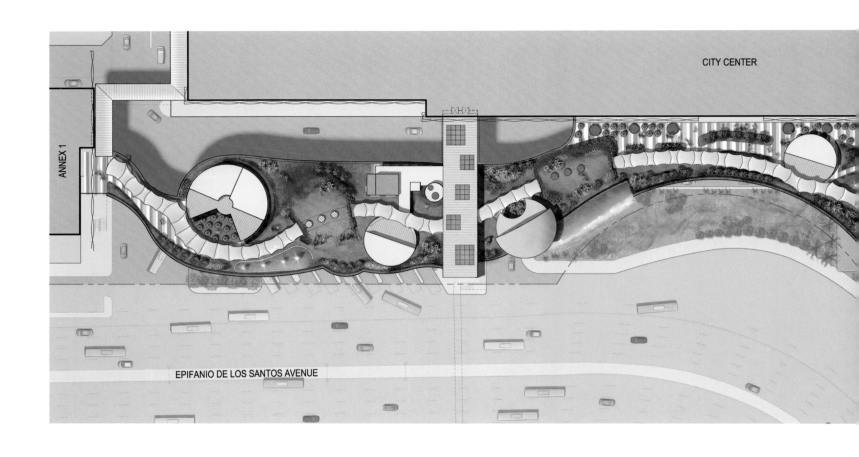

ANNEX 1

CITY CENTER

EPIFANIO DE LOS SANTOS AVENUE

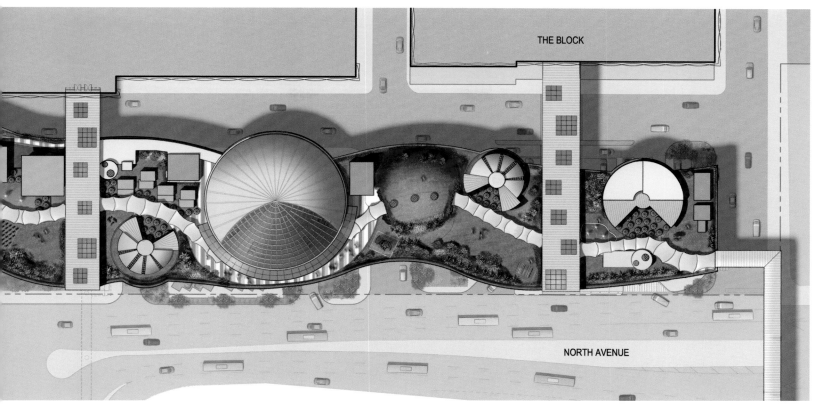

THE BLOCK

NORTH AVENUE

Masterplan 规划图

建筑环境和目标

"空中花园"为 SM 集团的第一个购物中心引入了绿色的主题。402 米长的露天花园成为一个绿色港湾，在这里，购物者和游客可以沿着绿化景观和覆盖有茂密的热带树木、开花植物和各种灌木的道路漫步。空中花园从购物中心的一端延伸到另一端，并且设计了大量的出入口通向街区、城市中心（主要购物中心）和附属建筑。实际上，这个花园是连接室外与室内的公共循环回路。当它全部完工的时候，将极大地增加购物中心的吸引力，同时这个设计理念也反映了全世界对于可持续性设计日渐增长的兴趣。

规划和设计概念

花园项目包括水池和喷泉、一个为演出和特别活动设置的露天剧场、餐饮大厅、休闲用餐地点、购物长廊和咖啡馆。一个落到地面的瀑布为设计方案增加了彩色景观，尤其当夜晚灯光点亮的时候效果更佳。此外，在每个拐角处和人行道的转角处设计有一些雕塑艺术品和互动小品，形成一种视觉享受。在空中花园下面是一个公共运输仓库和为购物者准备的附加停车区。

圆形活动区和餐饮大厅与购物中心圆形的天窗一致，而波浪起伏的楼板和人行道是外立面覆盖层和附属建筑缎带状活泼装饰物的延伸。当所有的设计元素加在一起时，建筑引人注目的设计不仅仅会将它与 SM 集团的其他购物中心区分开来，而且会使其成为一座标志性建筑，在比较中明显超越其他类似建筑。

景观策略

建筑的概念设计是将现有的购物中心与一个室外步行购物、餐饮中心连接起来，从而使其在结构、外形和色彩上与室外建筑相协调。水被巧妙地用做重要的元素来缓冲附近交通的噪声以及为整个项目提供感官上的丰富度。

建筑景观将会被绿植高密度地覆盖。花园也因此与项目的餐饮区域紧密地连接在一起，在创造公共空间享受方面发挥了重要的作用。

景区的生命力通过空间建立了起来，在这里人和花园交融在一起，绿色植被区域和圆形广场表演空间组成了总体的户外体验区。

在详细地调查行人的步行体验后，人行步道和隧道被用来创造户外购物和餐饮的环境，空间在传达品质和舒适的公园环境中从比较私密的空间向开放的空间转变。

总结

随着 SM 集团北部购物中心的全面发展，空中花园的设计在高效性、功能性、灵活性和近些年来被人广为关注的可持续性设计等方面满足了客户的需求。综合考虑了街区、城市中心和附属建筑，城市花园正逐步成为现代大都市游客的购物目的地。

NEW WORLD
K11 MALL WUHAN
新世界武汉 K11 购物中心

Architects: DCI
Location: Wuhan, China
Area: 25,000 m²
Client: Hong Kong New World Group

设计机构：美国 DCI 思亚国际设计集团
项目地点：中国武汉市
面积：25 000 平方米
客户：香港新世界集团

K11 is a brand of art shopping mall, subordinate to Hong Kong New World Group. It is the first innovative business model to integrate the three core elements "art, humanity and nature" with business. In 2009, K11 brand bloomed for the first time in Hong Kong and drew a lot of attention because of its innovative idea. In 2010, K11 was introduced into the mainland and the mainland trip of K11 started from Wuhan.

Brand strategies

K11 multicultural residential area is located in the downtown of Hankou, Wuhan, with K11 Avenue as its axis connecting New World Center, New World Department Store, New World Hotel, K11 Art Shopping Mall, chi K11 Art Space, City Farm, Vertical Green Wall and Humanity Graffiti Wall etc., creating a unique and special "Wuhan K11 Multicultural Residential Area".

K11 Art Shopping Mall is in the west of New World Center. Phase I development is "Gourmet Tower" which caters for various needs from banquet to date and fast food. As the future most distinctive food center in Wuhan, it accommodates many different food services, ranging from Chinese food, Western food and Western fast food to business meal, etc.

K11 Gourmet Tower has seven floors, five above ground and the other two underground. All kinds of foods you can find all over the world are served within the 20, 000 m² building. The unique decoration and show space endow it with multiple attributes of arts and humanities so that visitors can appreciate ubiquitous contemporary art while enjoying their sumptuous meal. Besides shopping, it is also an ideal place for social intercourse, leisure and rest. This project along with neighboring hotel, department store and future K11 Shopping Center forms the most splendid core business district in Wuhan.

From our perspective, successful commercial design is to provide interesting consumption experience and acts as a medium communicating brand ideology to the general public. Designer plays the part of disseminator and communicator. In the deeper level, based on the state elaborated by the brand K11, i.e., correlation between art and surrounding human life, such a typical urban retail building is supposed to sort and integrate surrounding human life in multi-dimensions to revitalize, retransform and to recreate the settled humanity art and life culture, creating a unique multi-cultural residential area.

K11, known as "Shopping Art Museum", overturns the traditional operation mode of present shopping centers. Within the unique area of K11, multi-dimensional art appreciation and communication, combined with retransformation and reproduction of local humanity, green architecture space, people's daily life and its atmosphere in the shopping mall and public area, produce subtle interactive chemical effect and provides the public with unprecedented sensory experiences.

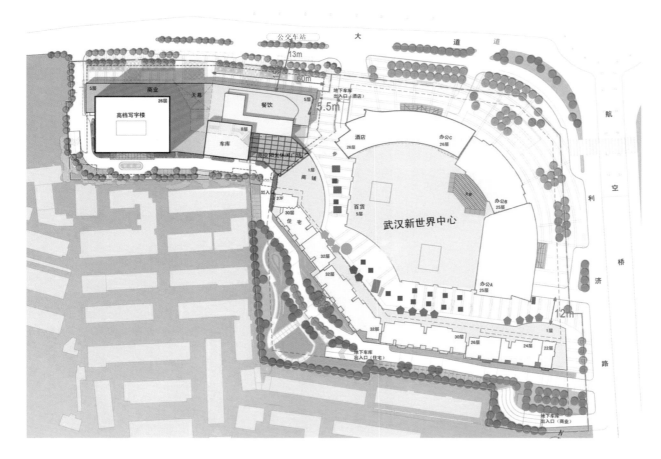

前广场面积为 900m²;
后广场面积为 650m²;
步行街全长为 270m

Plan 平面图

K11是香港新世界集团旗下的艺术购物中心品牌，是首个把"艺术·人文·自然"三大核心元素与商业相融合的创新性商业模式。2009 年，K11 品牌首次在香港绽放，因其创新理念而备受关注。2010 年，新世界集团将 K11 品牌引进中国内地，在武汉开始了 K11 内地的品牌之旅。

K11 品牌战略

K11 多元文化生活区位于武汉汉口中心，以 K11 大道为轴线贯穿新世界中心、新世界百货商场、新世界酒店、K11 艺术购物中心、chi K11 艺术空间、都市农庄及垂直绿化墙、人文涂鸦墙等，缔造出独特而充满个性的"武汉 K11 多元文化生活区"。

K11 艺术购物中心位于新世界中心西侧。它的一期开发项目定位为"Gourmet Tower 新食艺"，作为武汉未来最具有特色的美食中心，它汇聚中餐、西餐、西式快餐、商务正餐等多种饮食业态，可满足从宴请、约会到快餐等多样化的需求。

K11 新食艺共七层，地上五层，地下两层，在 20 000 平方米的建筑空间内世界美食应有尽有。同时这里独特的装饰艺术和展示空间赋予其人文艺术的多重属性，让人们在享受美食的饕餮盛宴之时，还可以欣赏到无处不在的当代艺术。而在购物消费之余，这里也是人们社交、闲暇、休憩的理想之处。项目与相邻的酒店、百货商场及未来的 K11 购物中心形成武汉最繁华的核心商业区。

在我们看来，成功的商业设计为人们带来有趣的消费体验，也是一种把品牌理念传递给大众的媒介，设计师在消费大众和品牌之间充当着沟通与传播的角色。在更深层次上，K11 品牌阐述了艺术与周边区域的人文生活之间的相互关系，对于这样一种典型的城市商业建筑空间，其最终使命应对所在的周边区域的人文生活进行多维度的梳理与整合，使该区域所沉淀的人文艺术及生活文化得到活化（"Revitalize"）、重塑（"Retransform"）和再生（"Recreate"），创造出独特的多元文化生活区。

而以"购物艺术馆"著称的 K11 正是对当前购物中心传统运营模式的一种颠覆。在 K11 所在的独特区域糅合多维艺术形式的欣赏交流、本土人文的重塑及再现、自然环保的建筑空间、商场及公共空间里人们的日常生活与气息，从而产生微妙的互动化学作用，为大众带来前所未有的独特感官体验。

TITANIC BELFAST
贝尔法斯特铁达尼号

Architects: CivicArts, Eric R Kuhne & Associates
Location: Belfast, UK
Area: 12,000 m²
Photographer: CivicArts

设计机构：CivicArts, Eric R Kuhne & Associates
项目地点：英国贝尔法斯特市
面积：12 000 平方米
摄影：CivicArts

Delivering the world's largest Titanic visitor experience, Titanic Belfast opened its door to the world on 31st March, 2012. The world's largest ever Titanic-themed visitor attraction and Northern Ireland's largest tourism project, Titanic Belfast is the result of a successful collaboration between the concept design architects CivicArts, Eric R Kuhne & Associates and the lead consultant architect Todd Architects.

Located in Belfast, Northern Ireland, on the site where the famous ship was designed and built, Titanic Belfast's six floors feature nine interpretive and interactive galleries designed by Event Communications that explore the sights, sounds, smells and stories of Titanic, as well as the city and people that crafted her, the passengers who sailed on her and the scientists who found her.

Event Communications' Chief Executive James Alexander explains: "this is a wonderfully exciting project for Belfast that not only tells the story of the Titanic, giving visitors the opportunity to look behind the scenes and marvel at the scale of Belfast's innovation and industry, but also seeks to dispel some of the myths and legends about the tragedy."

The building also houses temporary exhibition space, a 1,000-seat banqueting hall, education and community facilities, catering and retail space and a basement car park. Titanic Belfast has a complicated geometry, providing a challenging build programme which requires ground-breaking construction techniques. Its stand-out exterior facade, which replicates four 27 m-high hulls, is clad in 3,000 individual silver anodized aluminium shards, of which two-thirds are unique in design. The resolution of the geometries involved required the use of sophisticated 3D-modelling, completed by Todds in-house, in a process of "virtual prototyping" which was developed specifically for the project.

Titanic Belfast also incorporates the best design and technology available. For instance, the building adopted an integrated design approach in line with the Intergovernmental Panel on Climate Change Working Group III Guidelines and is on course for a BREEAM Excellent status. Plus, like Titanic, the project was completed on budget and to a strenuous time constraint which demanded completion in advance of the forthcoming centenary of the Titanic's maiden voyage in April, 2012.

This is a landmark development of Northern Ireland which we believe it will demonstrate the ability of iconic architecture to shape internal and external perceptions. Belfast has come far in the past 15 years and the landmark building such as Titanic Belfast reflects and reinforces the city's renewed sense of civic pride and cohesion.

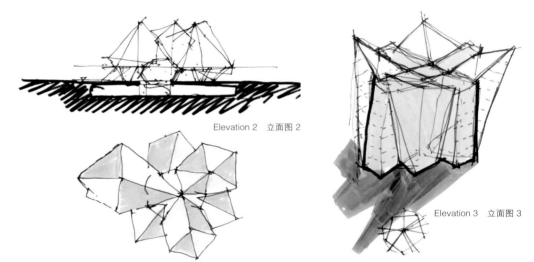

Elevation 2　立面图 2

Elevation 3　立面图 3

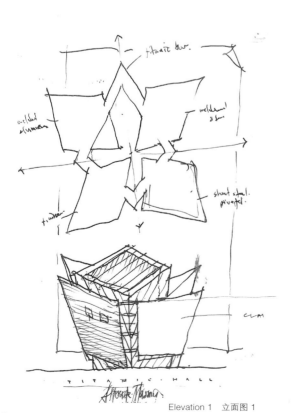

Elevation 1　立面图 1

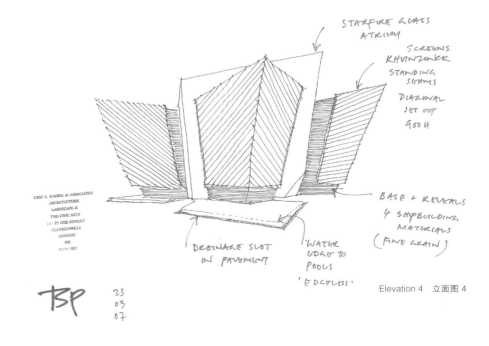

Elevation 4　立面图 4

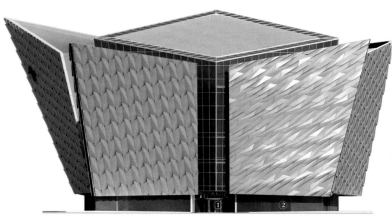

South Elevation 南立面图

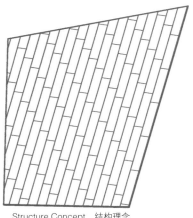

Structure Concept 结构理念

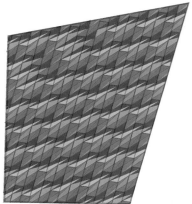

Cladding system 围护体系

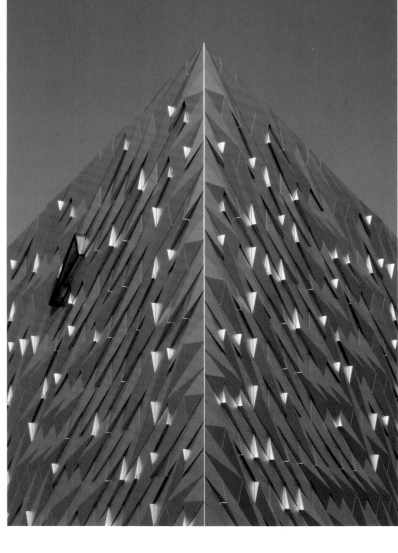

Typical 30 deg pattern 典型的 30 度模式

The backside with stucture exposed(in yellow)
结构外露的背侧（黄色部分）

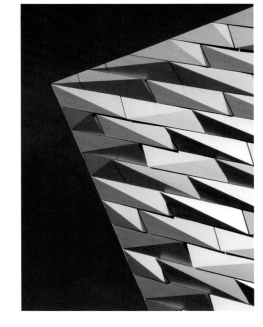

Facade system concept 表皮系统概念

4 basic modules 4 个基本模块

2012年3月31日，铁达尼号贝尔法斯特游客中心正式对外开放。该中心由CivicArts，Eric R Kuhne & Associates与Todd建筑事务所联合设计，是目前全球规模最大的以铁达尼号为主题的公园以及北爱尔兰最大的旅游项目。

本建筑位于铁达尼号这座旷世巨轮被建造和设计的旧址——北爱尔兰贝尔法斯特市。在这座6层建筑中由Event Communications公司设计了9个互动展厅，利用视觉、听觉、味觉等多种方式向人们呈现和铁达尼号有关的故事，包括建造它的城市和市民、当时乘坐它的乘客以及后来发现它的科学家。

Event Communications公司的首席执行官James Alexander介绍道："本项目对于贝尔法斯特来说是宏伟的，也是令人激动的。它不仅仅讲述了铁达尼号的故事，向游客展示了贝尔法斯特当时创新和工业幕后的奇迹故事，而且可以驱散关于这场悲剧的神秘感和传说。"

本建筑还设有临时展示空间、能容纳1 000人的宴会厅、教育和社区设施、餐饮和零售空间以及地下停车场。铁达尼号贝尔法斯特游客中心具有复杂的几何形体，这给施工带来了巨大的挑战，需要突破性的技术支持。4个27米高像船体一样突出的外墙被3 000个独特的镀锌银化铝片包裹，这些铝片中有三分之二形式独特。为了完成建筑形体复杂的三维建模，Todd建筑事务所专门开发了一种"虚拟原型"建模方法。

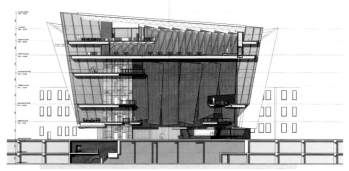

Section 1　剖面图1

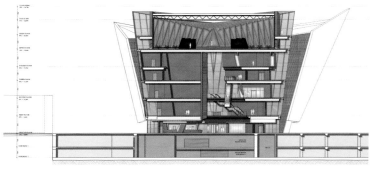

Section 2　剖面图2

Section 3　剖面图3

Section 4　剖面图4

　　铁达尼号贝尔法斯特游客中心结合了现有的最佳设计方法和建筑技术，例如建筑采用了《政府气候变化专门委员会第三工作组指南》推荐的综合设计方法，并且正在申请建筑研究所环境评估体系（BREEAM）的卓越等级。在有限的预算和紧张的建设周期中，建筑终于在 2012 年 4 月，也就是铁达尼号处女航 100 周年纪念这一天对公众开放。

　　这是北爱尔兰地区具有里程碑意义的标志性建筑，我们相信它具有完美诠释一座标志性建筑内部和外部特质的能力。贝尔法斯特在过去的 15 年里有很大的发展，像铁达尼号贝尔法斯特游客中心这样的标志性建筑，一定会增强贝尔法斯特市公民的自豪感和凝聚力。

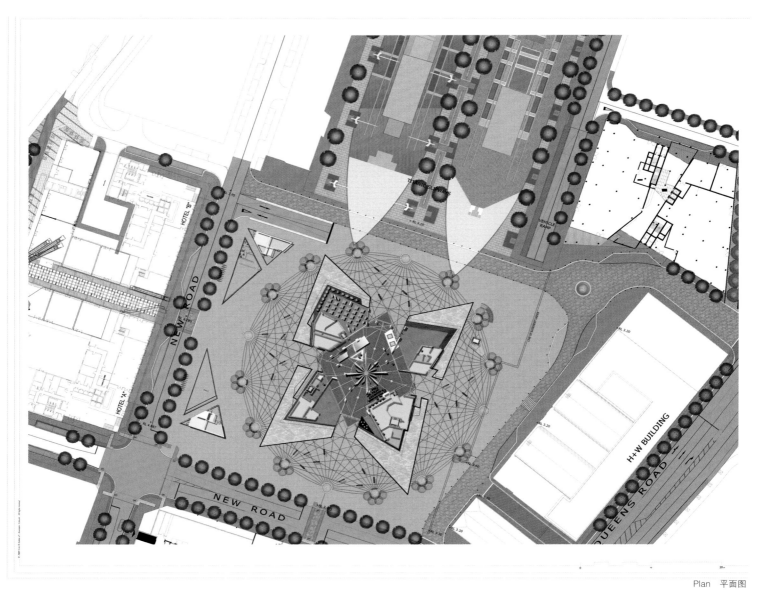

Plan 平面图

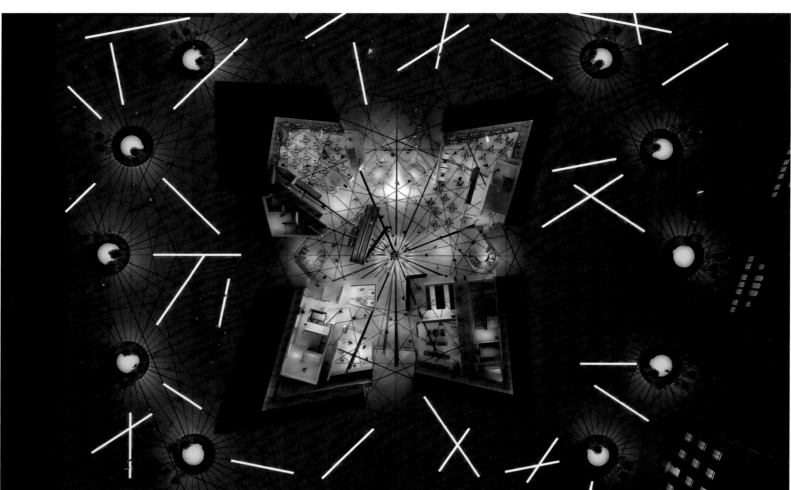

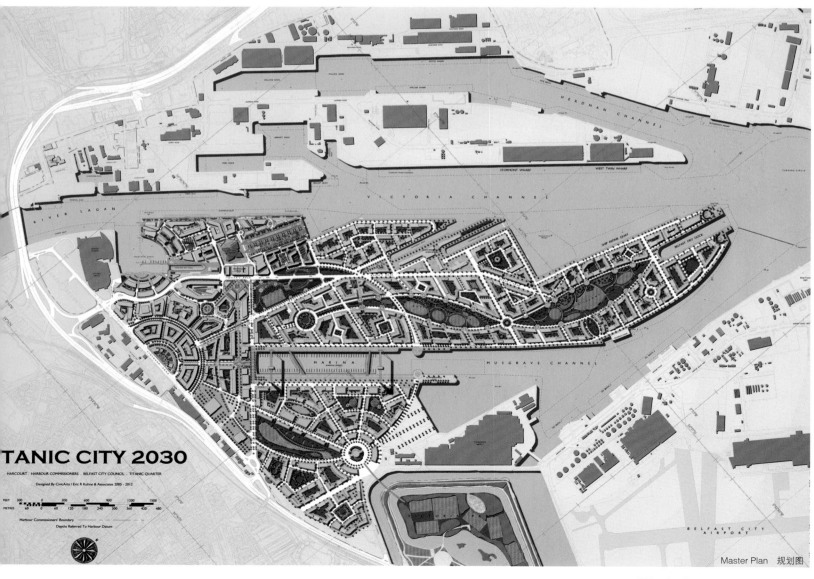

TANIC CITY 2030

HARCOURT . HARBOUR COMMISSIONERS . BELFAST CITY COUNCIL . TITANIC QUARTER

Designed By CivicArts / Eric R Kuhne & Associates 2005 - 2012

Harbour Commissioners' Boundary
Depths Referred To Harbour Datum

Master Plan 规划图

办公建筑 OFFICE BUILDING

Office building nowadays occupies a pretty high amount only second to residential house. Since we have entered a wholly new era of information due to the endless stream of emerging high-tech in the 21st century. The development of office building, as one of the main places to collect and handle information, is beyond comparison. According to an estimate, over a half of our population will work in office buildings in the middle of this century. So from this point of view, 21st century is "the century of office building".

As a kind of place to collect, handle and generate various administrative, scientific and commercial information, office building belongs to the infrastructure construction for social reproduction. It is a kind of place where information is produced, as well a place where we live. So the comfort level of the office directly affects the working efficiency of the staff. That is the reason why the requirement of a humanized design of a modern office building is exalted. Hence it is demanded that the design of the inner space of an office building should be as favorable to staff's working mind and behavior as possible. Similarly, a balance between the elegant, cozy and humanized outer and inner spatial environment is required. All these will greatly enhance the physical and psychological development of the staff, as well as the working efficiency.

Meanwhile, the spatial construction of office building is varying to adapt new official functions and new requirements for a higher comfort level. It is not only a request for a reasonable schedule of traffic routes, but also a result of spatial organization with a general consideration of elements such as aesthetics and technique. A mature and independent development system of office building construction is now gradually taking shape. Architects of office building are continuously seeking for brand-new design methods and architectural styles to capture the spirit of the era, and to create office works with the most era features and modern science and techniques.

It is gradually an important premise of architectural design to develop sustainably, environment- friendly and energy-savingly. How to maximize the function without the negligence of saving energy, and to make it a goal to be environment-protecting are among the key considerations of the architects. At present, the status of office building in the city is getting more and more important as its quality and quantity are increasing. It has become a significant part of the skyline of the city. Some momentous office buildings are even considered as landmarks of the city.

　　目前办公建筑已成为除住宅外数量外最多的一类建筑。在高新技术层出不穷的21世纪，人们进入了崭新的信息时代，办公建筑作为收集和处理信息的主要场所之一，其在信息时代的发展不可同日而语。据预测，到21世纪中叶，将有一半以上的人口在办公建筑里工作。从某种角度来讲，21世纪可谓是"办公楼的世纪"。

　　作为收集、处理和产生各种行政、科研、商务信息的场所，办公建筑是社会再生产的基础性建筑。办公室既是信息生产的场所，又是人们生活的场地，其环境的舒适度直接影响着员工的办公效率，因此，当代办公建筑设计对办公场所的人性化要求也逐渐提高，这就要求办公建筑内部空间的环境向着最有利于人员办公心理和行为的方向发展，同时要兼顾内外空间环境的美观、舒适，充满人性关怀，这些对员工身心健康的发展和工作效率的提高都能起到重要的促进作用。

　　同时，办公建筑的空间构成也在不断适应新的办公功能，随舒适性要求的变化而变化，这不仅是合理安排交通路线的要求，也是综合考虑审美、技术等多种因素后的空间组合的结果。办公建筑已逐渐形成了成熟而独立的发展体系，建筑师在不断寻求新的设计方法、建筑形式，以捕捉时代精神，创造出最具有时代特色、能够显示现代科技的办公建筑作品。

　　可持续发展和绿色环保节能理念逐渐成为建筑设计的重要前提，如何在实现功能最大化的同时节约能源、实现绿色环保的目标，成为建筑师在设计办公建筑时考虑的重点之一。如今，随着办公建筑数量和品质的增长，它在城市中的地位也愈加重要，开始成为城市天际线的重要组成部分；甚至一些重要的办公建筑，已作为城市地标而成为城市的象征。

TWIN TREE TOWER
孪生树塔

Architects: BCHO Architects Associates, Cho Byoung-soo/ Principle
Location: Seoul, Korea
Area: 54,918.85 m²
Photographer: Woo-seop Hwang, Kyoung-sub Shin, Yongk-wan Kim

设计机构：BCHO 建筑师事务所，Cho Byoung-soo/ Principle
项目地点：韩国首尔市
面积：54 918.85 平方米
摄影：Woo-seop Hwang, Kyoung-sub Shin,
Yongk-wan Kim

The city consists of an assortment of natural, humanistic and artificial flows. Flows of people, water, traffic, wind and lines generate a distinctive urban energy. It is this unique urban energy and site conditions that Twin Tree Tower strives to physically encapsulate, showing the concentration of powers.

According to the themes discussed in earlier projects, the two towers has been developed on the basis of three-dimensional representation of the irregular site conditions, allowing the public to physically experience the curvature of the street. The distinct form references the configuration of an age-old tree species in Korea—birch and its inherently strong structure is able to withstand the harshest climatic conditions. The tensions between the towers and vertical spaces, the building gaps and micro scales show the strong relationships. The layout of the space between the two buildings is special: the one stands forward slightly, while the other recedes subtly, which creating an in-between space provides a public gathering area. From the initial sketch, the facade remains within the confines of the site property lines, so this free-form surface has to be refined into pure arcs and lines. The undulating surface needs to be broken down into component levels to help determine construction assemblies. By analyzing these simple horizontal sections, we can define the basic form of the building: the top, bottom and middle sections are composed of circles and tangents. In turn, these lines create a geometry that can be constructed with standardized modules.

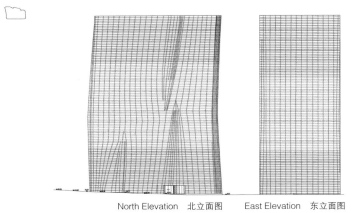

North Elevation　北立面图　　East Elevation　东立面图

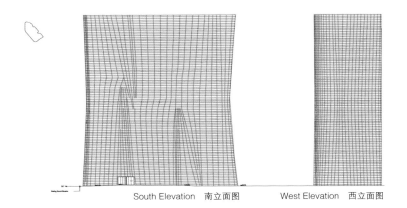

South Elevation　南立面图　　West Elevation　西立面图

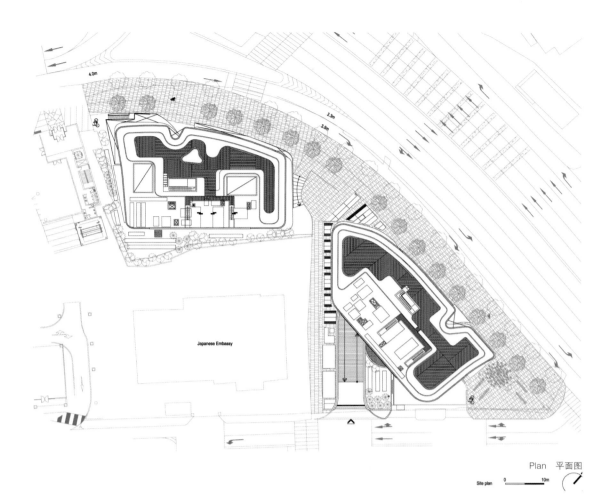

Plan 平面图

Site plan   0   10m

Although the exterior of the Twin Tree Tower appears to be simple, it is a quite complicated building comprised of numerous differentiated curves. It was particularly important to find a system to simplify it. The undulating curved glass and aluminum mullion system define the interior and exterior of the building, modulating views and reflections of the city. The shadows of these meandering lines fall into the office space early in the morning and at dusk. The depth of the horizontal protruding mullion was initially designed to be 210 mm and clad in light-absorbing zinc. This surface texture on the horizontal transoms can complement the luminance of the glass facade. The light-absorbing zinc was chosen to give the building a much more primitive and solid feeling to the people who walk on the sidewalk nearby and in-between the two buildings. The prestressed tension system was adopted to reduce the floor height. It is a structure that overcomes the inherent weakness of concrete and can create a longer span than the existing reinforced concrete structure. As wires and concrete are adhered closely, it is also structurally stable. The 25 cm thick plate is supported by a flat plate structure where no beams is presented for more than 10 m.

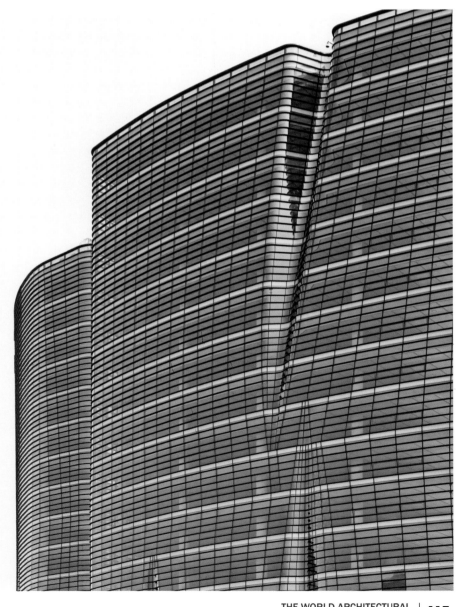

城市是由各式各样的、自然的、人文的和人造的流构成的。流动的人群、水、车辆、风和线条产生了独特的城市能量，正是由于这种独特的城市能量和场地条件，孪生树塔在体形上被压缩，彰显了力量的汇聚。

借鉴以往项目讨论的主题，两个塔楼根据场地不规则形状的三维表现而开发，允许大众体验场地街道的弯曲。建筑的独特形式借鉴了韩国古老桦树的外形，它强大的内在结构形式允许其承受巨大的风载荷和极端的天气情况。塔楼和垂直空间、建筑空隙以及城市微观尺度的张力展示着它们之间的强烈关系。两座建筑之间的空间构成独特：一座建筑稍微靠前，而另一座建筑巧妙地退向中心区域，形成了一个中间区域，提供出一个公共聚集区。从最初的草图开始，建筑的立面设计始终沿着规定的建筑红线，这种自由形式的表面必须被拆分成弧线和直线——连绵起伏的表面需要被拆分成各个组分，以便帮助确定施工部件的装配。通过分析这些简单的水平界面，我们可以确定建筑的基本形式：顶部、底部和中间截面由圆和其切线构成。反过来，这些线条创建的几何形体可以使用标准模块进行创建。

虽然孪生树塔的外观很简单，但实际上它是由众多差

Free form surface must be refined into pure arcs and lines. To achieve this we first create section lines of the n.u.r.b.s. surface

By analyzing these sample sections we can see the basic form of the building that can be defined by the top, bottom, and the middle sections.

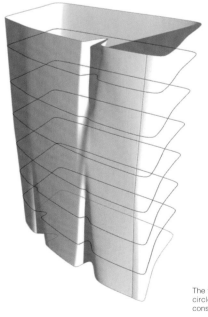

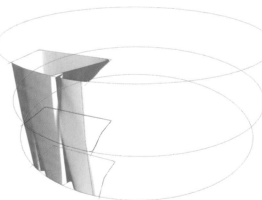

The standardized curves approximate the original form

The top, bottom, and the middle sections are traced with circles and lines. This creates a geometry that can be constructed with standardized modules.

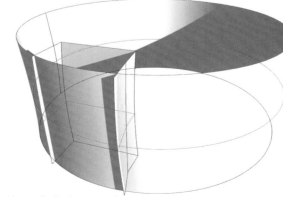

"The name lofting comes from the shipbuilding industry. It was found that work done in the mold loft served a useful function in constructions where parts with complex form had to be fitted together to build a ship. The definition of the hull shape was done in the loft over the shipyard, using enormous drawings. To provide a smooth longitudinal contour, points taken from desired cross sections were connected longitudinally on the drawing using flexible lead "ducks" (weights holding down a flexible rule).

Source:
LOFTING AND CONICS IN THE DESIGN OF AIRCRAFT

By lofting* the standardized curves a surface is created that approximates the origina.

Analysis 1　分析图 1

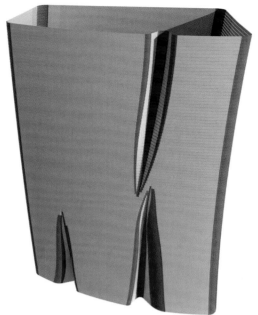

The red areas illustrate areas of curved glass in the facade

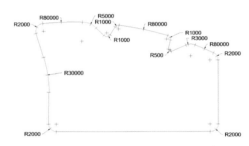

Analysis 2　分析图 2

Measurements of a section generated from the "simplified" surface verify that the radii have remained constant from the bottom of the building to the top.

异化曲线组成的复杂建筑，因此找到一个简单的系统简化它十分重要。波浪状的弧形玻璃和铝制窗框系统定义了建筑的内部和外部空间，调整了整个城市的视野和印象，使这些弯折线的阴影在清晨和傍晚落入办公空间。水平突出的窗棂的深度最初设计为 210 毫米，并且涂有吸光的锌材料。经过多次尝试，水平横梁的表面纹理可以补充玻璃幕墙的亮度。吸光的锌材料可以使人们在两座建筑附近和之间的人行步道穿行的时候，感受到建筑的复古感和坚实感。建筑使用预应力张拉弦系统来降低楼板的高度，该系统可以克服混凝土固有的缺点，能比现有的钢筋混凝土结构形成更大的跨度。由于绞线和混凝土黏结得十分紧密，该系统在结构上是稳定的。25 厘米的厚板由平板结构支撑，该体系在 10 米内都没有梁出现。

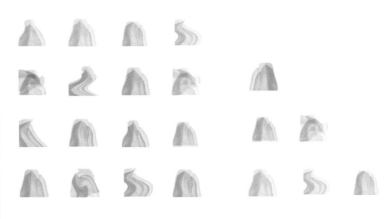

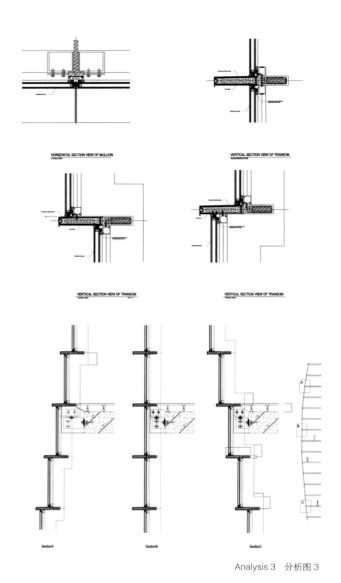

BAHRAIN WORLD TRADE CENTER
巴林世界贸易中心

Architects: Atkins Middle-East
Location: Manama, Bahrain
Area: 120,000 m²

设计机构：Atkins 中东公司
项目地点：巴林麦纳麦市
面积：120 000 平方米

The Bahrain World Trade Center stands as an icon of sustainable design engineering. It is the first building that installed large-scale wind turbines.
Atkins was commissioned to develop a master plan for the prominent site on the King Faisal highway in Manama. More than half of the area was previously developed and included a hotel, single storey mall, office tower, car parks and landscaped areas.
Our project's outstanding development location offers unobstructed views over the Arabian Gulf; we saw the opportunity to develop a landmark for Bahrain and to go above and beyond a standard rejuvenation plan. We not only wanted to provide high quality office accommodation but also to design something that was unique.
The Bahrain World Trade Center (BWTC) is the center piece of our master plan. With initial inspiration taken from traditional Arabian wind towers, our architects developed the idea that buildings could harness the prevailing onshore breeze from the gulf and if designed correctly, deliver a renewable source of energy for the project. Unique to this building and rising to the challenge of incorporating renewable energy solutions with sustainable architecture the design provides three 29 m diameter wind turbines to be horizontally supported between the two towers. The sail-shaped profiles of the two towers funnel the onshore breeze between them as well as creat lift behind, thus further accelerate the wind velocity between the twin structure. Tapering to a height of 240 m, each tower is visually anchored to the ground by a concertina of curved, sail-shaped forms. Atkins' architects and engineers undertook months of meticulous research, including extensive dialogue with specialist in turbine manufacture during the concept feasibility study and design development stages. Technical validation included the incorporation of environmentally responsive design elements, different wind regime analysis of turbine performance and SARM analysis validation. Results from the simulation modelling yielded the production of the final design.

Modelling　模型图

The designs were additionally supported in-house by our multi-disciplinary teams of:
• Structural engineering
• MEP
• Master planning
• Construction supervision
The turbines produce between 11%~15% of the total electrical consumption of the building. Since commencing construction, BWTC has been noted as one of the most distinguished and environmental friendly building designs to date. Its prominence in Manama has prompted moves of several major banks and legal firms to make BWTC as their local headquarters.
In 2007, the world's first integrated wind turbine power source was officially realised as the turbines were installed. The project was completed in April 2008 with the turbines officially certified by Bahrain's Electricity Distribution Directorate (EDD) in December 2008.

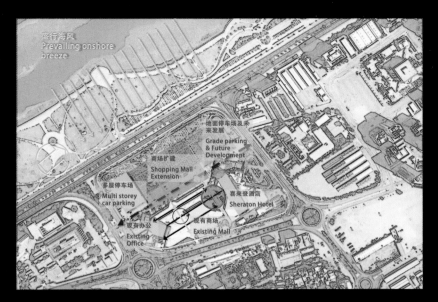

背景

Context
背景
Context

Forms the focal point of a
masterplan to rejuvenate an
existing hotel and shopping mall
on a prestigious site overlooking
the Arabian Gulf in the downtown
central business district of
Manama, Bahrain

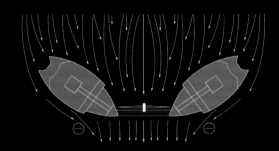

风涡轮
Wind turbines

The two 50-storey-sail-shaped
office towers taper to a height
of 240m and support three
29m diameter horizontal-axis
wind turbines

Elevation 1　立面图 1

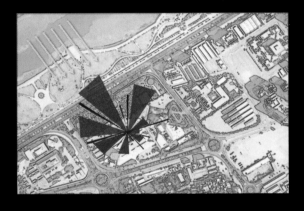

漏斗效应
Funneling effect of towers

The elliptical plan forms act as
aerofoils, funneling the onshore
breeze between them as well as
creating a negative pressure
behind, thus accelerating the wind
velocity between the two towers.

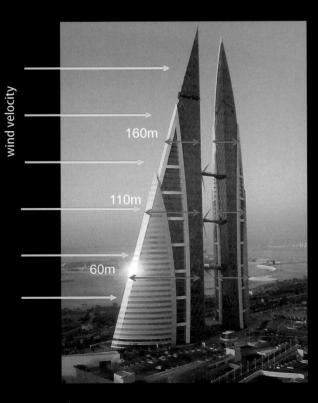

Onshore wind velocity
海风风速

160m
110m
60m

风动态
Wind Dynamics

Vertically, the tapering of the towers
reduces the funnelling effect at higher
levels. At the same time the wind
increases in velocity with height.

This, creates a near equal wind regime on
each of the three turbines.

This is a key factors that has allowed the
practical integration of wind turbines in a
commercial building design cost effective.

It resulted in
 Reduction in R&D
 Simplified risk analysis.
　Each turbine produces similar energy
　Turbines rotate at the same speed

Elevation 2　立面图 2

　　巴林世界贸易中心是可持续设计工程的标志性建筑，它首次将大规模风力涡轮机安装到商业建筑中。
　　Atkins公司被指定在麦纳麦费萨尔国王公路的显著景区上进行项目设计。该景区超过一半的面积曾经被开发过，包括一个酒店、单层购物中心、办公大楼、停车场和景观区。
　　项目优越的地理位置提供了阿拉伯海湾的畅通视角，我们看到了发展巴林地标性建筑和复兴计划的绝佳机会。我们不仅仅提供了高质量的办公住所，而且设计出了一个卓越的建筑项目。
　　巴林世界贸易中心(BWTC)是我们主体设计的核心部分。建筑的设计灵感来源于传统的阿拉伯风塔，通过正确的设计，可以使其驾驭海湾吹来的微风，为项目提供可再生能源。
　　建筑的独特之处和最大挑战是将可再生能源方案和可持续发展建筑结合起来，三个直径为29米的风力涡轮机被设置在两座塔楼之间的水平支撑上。
　　建筑风帆形状的外形使陆地上的微风在两个塔楼之间形成漏斗，并且在建筑后面形成风力梯度，加快了风在两座建筑之间前进的速度。建筑在240米的高度上逐渐变尖，每一个塔楼在视觉上被六角手风琴形状的曲线底座固定在地面上，外形上像一个风帆。Atkins的建筑师和工程师进行了数月细致的研究工作，与涡轮机制造专家在概念可行性研究和设计阶段进行广泛的对话。方案的技术验证包括环境响应设计元素的编入、不同风荷载下的涡轮机性能分析和异步动力响应验证分析，最终由仿真模拟的结果得到了设计方案。

Site plan　总平面图

Ground floor plan　底层平面图

此外，本项目的设计是在 Atkins 多个团队的支持下完成的，它们包括：

· 结构工程
· 建设工程
· 总体规划
· 施工管理

这些涡轮发电机为建筑提供的电力为总耗电量的 11% 到 15%。从刚刚建设到目前为止，巴林世界贸易中心已经成为最著名的环境友好型建筑。其在麦纳麦的卓越地位已经促使几个主要银行和法律公司将它们的总部搬到了这里。

在 2007 年，随着涡轮机的安装，世界上第一个综合风力涡轮机发电系统正式完成。本项目在 2008 年 4 月正式完工，涡轮发电机组在 2008 年 12 月正式通过巴林电力分配理事会（EDD）认证。

UNITED STATES FEDERAL COURTHOUSE
美国联邦法院

Architects: Adrian Smith, Gordon Gill Architecture
Location: Miami, Florida, USA
Area: 53,678 m²
Photographer: J Espana, Robin Hill, Norman McGrath

设计机构：Adrian Smith 和 Gordon Gil 建筑师事务所
项目地点：美国弗罗里达州迈阿密布
面积：53 678 平方米
摄影：J Espana,Robin Hill, Norman McGrath

The building stands at the end of the axis created by the 4th Street Promenade. However, because the center atrium space is approximately 20 meters above the plaza, both the physical and visual axes are maintained through the building. The design completes the judicial campus created by the old courthouse, the courthouse annex tower, the Lawrence King Building and the federal prison.

The planimetric design is based on a four-courtroom per floor layout. The two court towers are connected by a generous circulation lobby, which occurs on every floor, and rises dramatically throughout the entire height of the building. A dramatic conical glass atrium connects the floors by piercing through the lobbies and decreasing in size as it rises through the building terminating in a skylight.

The site plan is divided into two distinct zones. The first, at street level is comprised of densely planted native vegetation punctuated with seating and meandering walks which can be used for local art fairs and markets. The second zone is the lawn. It is raised approximately 1 meter above the sidewalk forming a natural and secure perimeter around the building. The walls of the plinth have seating. The two east quadrants of lawn areas were designed as the sites in which the artist Maya Lin developed her earthen installation, "Flutter". The hard-scape is primarily comprised of three-colored pre-cast pavers, which coordinate with the pre-cast panels of the building.

　　建筑坐落于第四街长廊形成的轴的末端。由于中心的中庭空间比广场高大约20米，所以建筑物的物理轴和视觉轴都穿过建筑。设计完成了由旧法院、法院附楼劳伦斯景大厦和联邦监狱组成的司法校园。

　　平面设计基于每个楼层四个法庭的布局。两幢法院大楼由圆形的大厅连接，这种布局每层都有。一个锥形的玻璃中庭穿过大堂连接楼层，由于其终止于天窗所以在上升的过程中尺寸逐渐变小。

　　总设计图分为两部分：第一部分，街道层，这里有密集种植的原生植被，休息区和蜿蜒的散步小径，可用于当地的艺术展览和集市；第二部分是草坪，草坪大约比人行路高 1 米，形成了自然景观，并且保证了建筑周边的安全。建筑底座区域可以用来休息。东边的两个扇形草坪区被艺术家 Maya Lin 设计成 "Flutter" 形式。这个硬质景观主要包含三色预制的铺路材料，与建筑的预制板相协调。

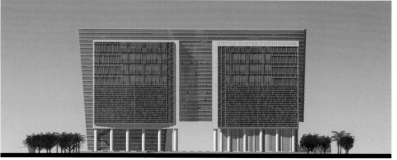

Elevation 1　立面图 1

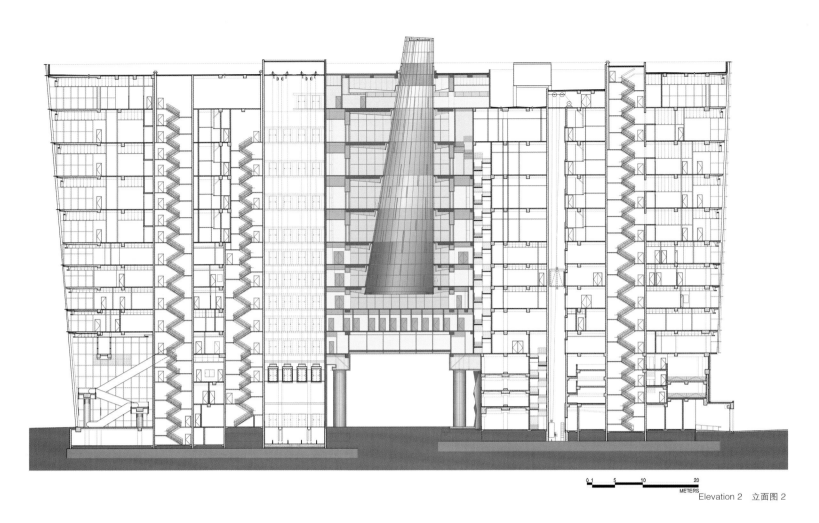

Elevation 2　立面图 2

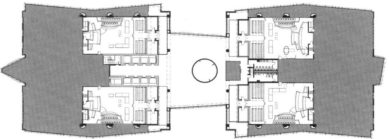

Courts Floor Plan　法院楼平面图

The building is composed of three individual elements: the opposing two towers and the glass crystal curtain-wall that spans between them.

The east and west curtain-wall elevations of the two towers are blueprints for the interior functions within the building. The alternating rhythms, depths and colors of the horizontal and vertical sunshades delineate the primary office, circulation, chamber and courtroom functions located in the tower.

The north and south facades of the tower are solid pre-cast with punched openings reinforcing the solidity of the envelope which then create the asymmetrical frame on the east and west elevations.

Signages are located on all four elevations of the top. At the top, it welcomes pedestrian, at the bottom it acts as a banner for the vehicular traveler. The exterior expression is primarily transparent and open reflecting the integrity of the court while emphasizing equality before the law.

A monumental three-color stone colonnade recalling the richness of traditional courthouses supports the south tower. The north tower sits upon the building garage clad in stone columns placed at irregular intervals and a metal grille. The crystal curtain-wall spanning the two towers contains an atrium that pierces through each floor.

　　该建筑由三个独立的单元构成，即两座相对的塔楼和一座横跨其间的玻璃水晶幕墙。
　　两座塔楼的东西幕墙立面为建筑内部功能描绘出了一幅美好的蓝图。交替变动的韵律，水平和垂直光影的深度和颜色突出了办公室、人员流通空间、会议室、法庭等功能设施。
　　塔楼的南北立面设置了牢固的带孔开口，加固了大楼外层，在东西两立面则形成了不对称的框架。
　　塔楼四面都有指示牌，底部的指示牌为来往的行人指引方向，顶部的指示牌像一面旗帜欢迎远道而来的客人。透明开放式的外观传达了一种法庭公正无私、法律面前人人平等的理念。
　　巨大的三色石柱廊支撑起南塔楼，让庄严的法院显得华美壮观。大楼车库上方坐落着不等距排列的石柱廊和金属栅栏支撑起的北塔楼。水晶幕墙横跨两座塔楼，塔楼中庭穿过每一层楼。

ALMAS TOWER
DUBAI, UAE
阿联酋迪拜 Almas 大厦

Architects: Atkins Middle-East
Location: Dubai, UAE
Area: 183,000 m²
Client: Dubai Multi Commodities Centre

设计机构：Atkins 中东公司
项目地点：阿联酋迪拜
面积：183 000 平方米
客户：迪拜多功能商业中心

Atkins was commissioned to provide conceptual design, detailed architectural, structural, landscaping, vertical transportation and electromechanical design services as well as full site supervision for this 360 m high commercial tower. The building is the centre piece of the Jumeirah Lake development, 20 km south-west of Dubai, and is host to the diamond exchange facility. The tower comprises 60 commercial office floors with facilities including:
• A diamond exchange centre
• A retail mall
• Ballroom
• A health spa and pool
• Parking spaces for 1,800 vehicles
Almas Tower takes the shape of two intersecting ellipses in plan, spanning 64 m at maximum length and 42 m at maximum width. The lower tower faces north with the taller being south facing, overlapping along their east west axis. One of the greatest challenges was dealing with the stress imbalance caused by the fact that one part of the building finishes 60 m below the other. We therefore designed one end of the core thicker by 150 mm and monitored the placement of the floor slabs. If these were not level, this would have induced more vertical stresses.
To ensure the building remained aesthetically balanced, we carried out analysis based on total vertical loading, taking into account the concrete strength, the effects of increasing the thickness of some core walls and the effects of creep. The final expected deflection was found to be 170 mm and we did not want to deflect more than 0.001% of the building's height. The current movement has not exceeded 50mm.
The north tower has a semi-transparent elevation, thereby maximising the cool, ambient northern light; whereas the exterior of the south facing the north tower has a high performance finish to maximize protection from the heat.

The client had set an 80% efficiency objective for this 360 m tall tower, the tallest structure Atkins had designed at the time. To maximise the lettable floor space, the building was designed with large peripheral columns connected by 1 m deep, 500 mm wide beams to minimise the need for interior columns. We executed vertical transportation design reiterations reducing the elevator number requirement and stacking services on top of one another in order to save space, without compromising end user service.

33 lifts service the building, with super fast vertical transportation of up to seven metres per second.

"Almas," meaning "diamond" in Arabic, was the inspiration of the distinctive projecting facets of the two-storey steel podium at the base of this 60-storey tower. Eight diamond facets reach out in total with the most prominent facet housing the actual diamond exchange, the largest of its kind in the Middle East, where stones can be viewed and traded. The podium glass is of a specification such that the diamond inspection process will not be influenced by the light. A glass floor of over 15 m covers the tip of the exchange, allowing clients to walk on the glass with the lake beneath them.

With Almas Tower housing the diamond exchange, security design was critical. The actual exchange needed to integrate seamlessly with the building, and yet be isolated in terms of access. Atkins worked very closely with the special security consultant to ensure the design met the security brief.

An 81 m steel-framed lattice spire tops off the tower. It is connected to the building with a 21 m reinforced concrete upstand and is clad with aluminium. The spire tapers, so the higher it ascends the more likely it is to vibrate. Four 2T-tuned mass dampers are fitted to mitigate the excess vibration of the mast during construction, a vortex suppression device was temporarily fitted to the top of the spire.

The whole project was run by Atkins with construction management undertaken by Faithful & Gould, an Atkins, subsidiary.

Almas Tower will carry Dubai's diamond trading forward for decades and allow Dubai to remain firmly on the map as a trading destination hub.

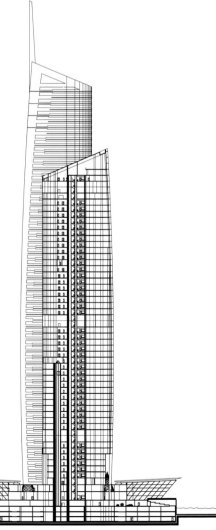

Elevation 立面图

Atkins 公司在 Almas 大厦的设计中提供了概念设计、详细建筑设计、结构设计、景观设计、竖向交通设计、机电设计及施工监管等服务。Almas 大厦是钻石交易中心，位于迪拜西南角的朱美拉湖发展区，是该区的标志性建筑。该大厦为 60 层的商务办公楼，包括如下设施：
　　·钻石交易中心
　　·商场
　　·宴会厅
　　·水疗区和泳池
　　·1 800 个停车位
　　Almas 大厦的典型平面是两个相互交叉的椭圆形，平面长 64 米，宽 42 米。其较低的塔楼朝北，较高的塔楼朝南，与建筑的东西轴相交叠。建造时结构上最大的挑战之一是如何处理建筑建造时的高低差（较高的塔楼比较低的塔楼高出 60 米）带来的压力不平衡。因此，我们建造的时候将核心的一端加厚 150 毫米，并监测楼板位置，如果不水平的话，将导致更大的垂直压力。
　　为了保持建筑美观，我们在分析整体竖向载荷的基础上，同时考虑了混凝土的强度，也考虑了增大一些核心墙的厚度是否带来消极的影响及是否有蠕变效应。偏转的最终期望值是 170 毫米，我们希望偏转不超过建筑物高度的 0.001%，目前运转尚未超过 50 毫米。
　　北塔采用半透明的外立面设计，从而可以尽最大可能地隔热及利用北边采光，达到最有效的低碳设计。南塔也采用隔热设计。
　　客户期望建造 360 米高的大厦，这也是目前 Atkins 设计的最高的建筑物。为了最大化各层出租面积，整个设计用 1 米深、500 毫米宽的梁连接建筑四周的承重柱，这样就可以尽量减少内柱的需要。我们设计垂直交通的时候，在不影响对用户的服务的前提下，减少电梯的数量，采用堆叠服务，节省空间。
　　整个大厦一共有 33 部电梯，提供最快的服务，速度基本达到 7 米/秒。

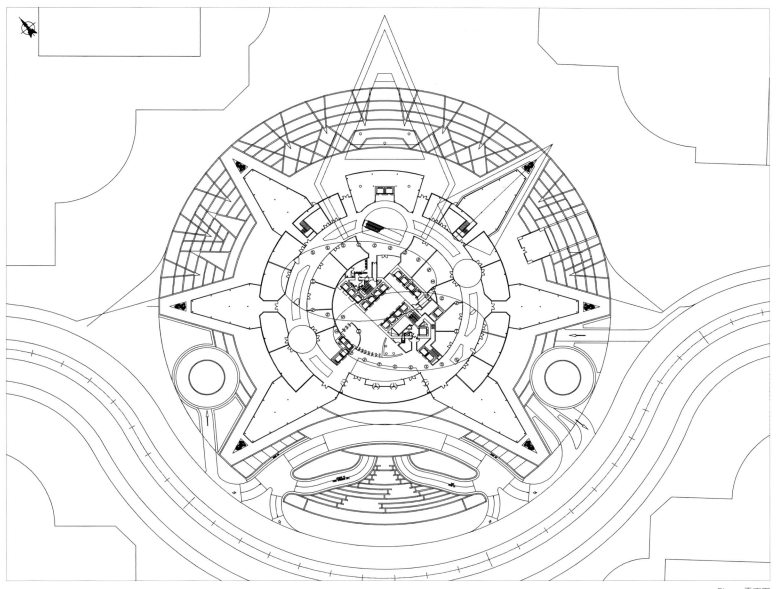

Plan 平面图

　　"Almas"在阿拉伯语里是钻石的意思，建筑底部2层极其特别，其灵感来自钻石的八个面。这里是中东最大的钻石交易中心，可以在这里观赏和交易钻石。裙房采用玻璃材质，使得它就像钻石一样闪闪发光。交易中心的地面用15米玻璃覆盖，行走在上面犹如踏步在湖上。

　　Almas 大厦的安全设计可想而知是至关重要的，整个钻石交易过程需要绝对的安全，所以对独立的安全主入口设计要求极高。因此 Atkins 团队和安全顾问团队紧密地合作，确保大厦的安全系数。

　　Atkins 大厦顶部设有一个81米的钢架晶格尖顶，它利用21米的钢筋混凝土柱连接着大厦，它的材质是铝包钢。由于尖顶逐渐变细，故越高震动越明显。在施工阶段，我们安装了4个2T调谐质量阻尼器，以减轻多余的震动；同时暂时在塔尖顶部安装了一个旋涡抑制设备。

　　阿特金斯是该项目的总承包公司，并由其子公司 F&G 进行施工管理。

　　Almas 大厦将推进迪拜的钻石交易，同时确保迪拜作为钻石交易枢纽站这一坚固稳定的地位。

AL SALAM TECOM TOWER DUBAI, UAE

阿联酋迪拜 Al Salam Tecom 大厦

Architects: Atkins
Location: Dubai,UAE
Area: 94,500 m²
Client: Abdulsalam M Rafi Al Rafi

设计机构：Atkins
项目地点：阿联酋迪拜
面积：94 500 平方米
客户：Abdulsalam M Rafi Al Rafi

The Al Salam Tecom Tower in the Dubai Technology, Electronic, Commerce and Media Free Zone on Sheikh Zayed Road stands at 195 m. This mixed-use building comprises a retail area on the ground and first floors, serviced apartments on the lower 15 floors and offices on the upper 23 floors.

Atkins' scope of services includes the conceptual design, detailed architectural, structural, landscaping and electromechanical designs, as well as full site supervision. The 47-storey building resting on a 28 m high podium accommodates recreational facilities at deck level, including two swimming pools and two health clubs—one individually dedicated for the serviced apartment tenants and the other for office space tenants. The ground floor comprises a retail area and the entrance lobby to the apartments and office area. The two basement levels and five podium levels are dedicated to parking spaces. The food court, business centre and further retail areas are located at first floor level.

The serviced apartments are accommodated on 15 typical floors and comprise a mix of studio, one-and two-bedroom apartments. The office floors are open planned and provide raised floor system to allow maximum flexibility of furnishing for the tenants. The most significant feature of the building is the inclined expression which emphasises the two triangular masses composing the tower's main elevation, behind which the office floors are located. The residential component is expressed by the use of precast panels with window treatment and balconies to highlight and separate the mixed-use elements of the building. Reflective silver and blue curtain glazing is used along with tinted grey glazing, helping to define areas of the structure. The use of simple forms contributes to achieving an elegant and hi-tech design.

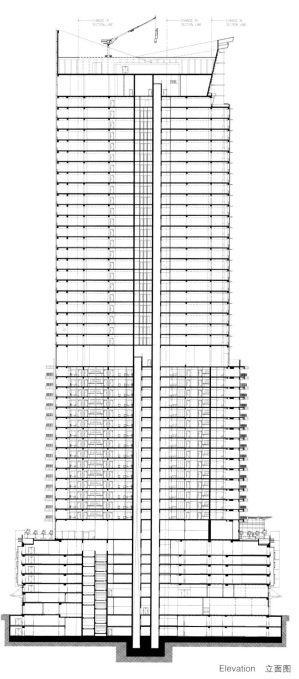

Elevation 立面图

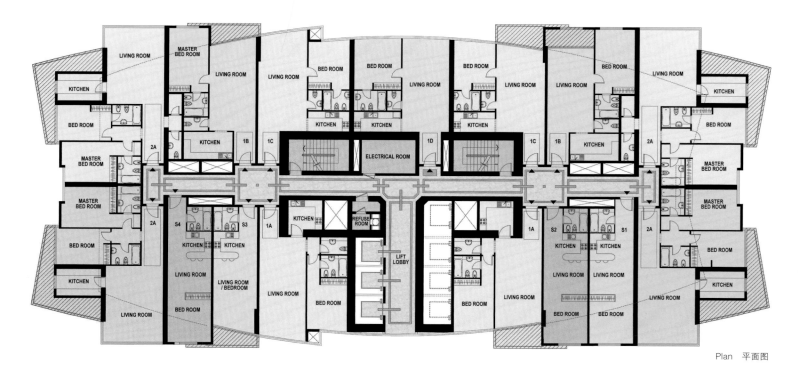

Plan 平面图

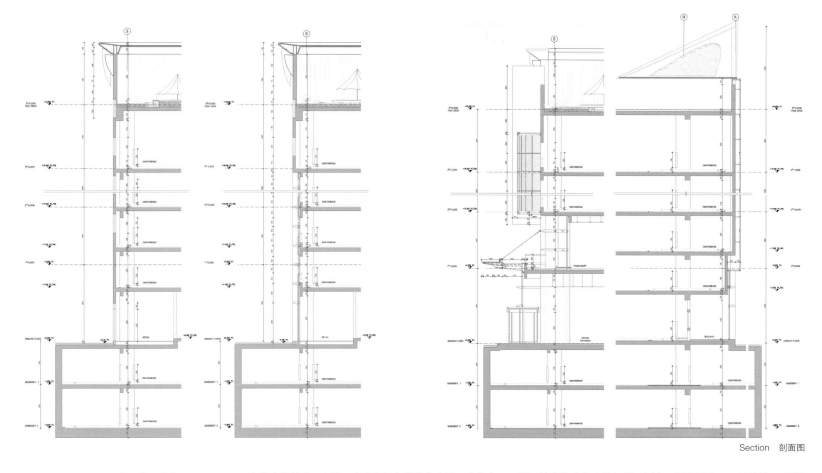

Section 剖面图

    Al Salam Tecom 大厦位于迪拜 Sheikh Zayed 大道上的技术、电子、商业贸易和媒体自由区，建筑高 195 米。这座综合体建筑包括 1 层和 2 层的零售区、低层的 15 层酒店式公寓和高层的 23 层办公区。

    Atkins 公司的服务范围包括概念设计、详细的建筑设计、结构设计、景观设计、机械电气设计以及全程场地监理服务。这座 47 层的大厦耸立在 28 米高的裙房之上，裙房的露天层设有娱乐设施，包括两个游泳池和两个健康俱乐部，其中一个为酒店式公寓的住户服务，而另一个为办公区的租户服务。建筑的一层由零售区和通向公寓和办公区域的门厅组成，两个地下室和五个平台层提供专用的停车区域。美食广场、商务中心和更多的零售区位于一层的位置。

    酒店式公寓占据了建筑的 15 层楼，由工作室、一居室和两居室的公寓混合而成。办公区域开敞布置，并提供了高架地板系统，为租户设计办公空间提供了最大的自由度。本建筑最显著的特征是建筑倾斜的外形：两个三角形块体构成了建筑的主要外立面，在它后面是主要的办公楼层。住宅部分使用带窗户和阳台的预制混凝土板来表达，突出和分离了本建筑的两种功能元素。具有反射面的银色和蓝色玻璃幕墙和灰色的玻璃搭配使用来界定不同功能的区域。建筑使用简单的形式实现了优雅和高科技并存的设计。

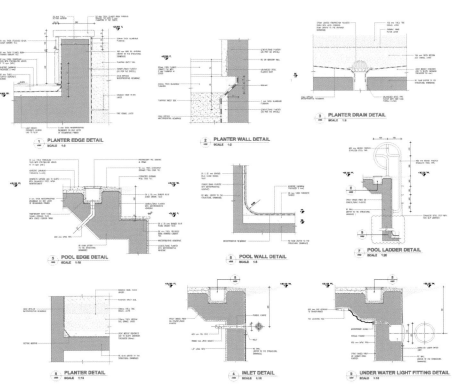

Analysis 分析图

BANK MUSCAT, OMAN
阿曼马斯喀特银行

Architects: Atkins
Location: Muscat ,Oman
Area: 31,000 m²
Client : Bank Muscat

设计机构：Atkins 公司
项目地点：阿曼马斯喀特市
面积：31 000 平方米
客户：马斯喀特银行

The building is a very contemporary design. The complex comprises of four L-shaped interconnecting buildings with courtyards. The interconnecting central spine is the ground floor "street", complete with restaurants, shops, crèche, gym and auditorium to create the big happy family atmosphere to the 2,000 Bank Muscat employees and its customers. This is a variation upon successful solutions used by other corporate companies and architects around the world. The most notable British Airways Headquarters building in London designed by architects Niels Torp and Potsdamer Platz with its internal street designed by the architect Renzo Piano is the inspiration. The central courtyard is a Moroccan Garden with water features and is a direct indoor-outdoor connection with the street. The two other courtyards are the main entrances to the building which are enhanced by natural light entering through tiered bands of glass, producing a floating effect to the high level atrium roofs.

Orientation and shading is vital in the design of building elevations in Oman and this has the greatest effect on solar gain. The development of the building elevations is to reduce its overall impact on the environment by incorporating several key design features, including facade design features and improvements over standard mechanical services design and electrical services design, as well as water saving features. This was done with the help of our regional sustainability team during the preliminary design stage of the project. North and south facades are relatively easy to shade (using window overhangs), while east and west facades are subject to low angle sun light in the morning and afternoon by the use of suspended sash windows . The building is oriented 28° west of north (nearly west-north-west) and the full height glazed (floor to ceiling glass) curtain walling facades were introduced to north and south elevations, making them easy to shade. Shading was done with the introduction of aluminium screens replicating an abstract visual aesthetic of a traditional Islamic pattern. On east and west facades the windows were recessed in the wall and horizontal overhangs were introduced. Importantly, these windows are also reduced in size compared to the ones on north and south, with the overall effect of solar gains and radiant temperatures being greatly reduced. The building elevations are also created to reinterpretate the cultural context and tradition in a contemporary way.

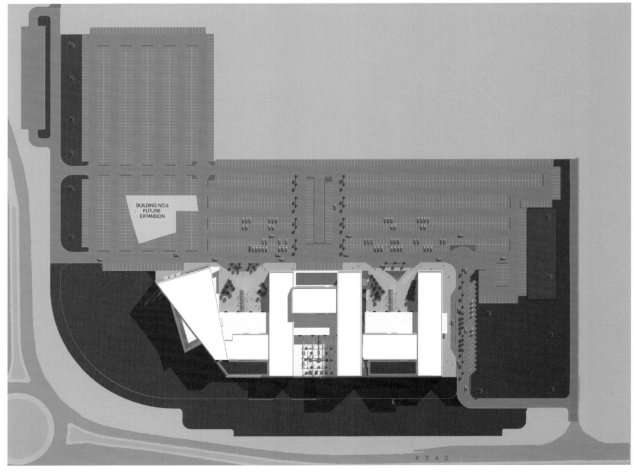

Plan 平面图

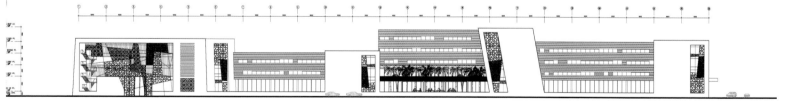

Section 剖面图

　　本建筑是一个非常现代的设计，包括带有中庭的四个相互连接的"L"形建筑。一层的大街像脊柱一样将各个建筑连接起来。建筑设有餐馆、商店、托儿所、健身房和礼堂，为马斯喀特银行的 2 000 名雇员和顾客创造出一种温暖的大家庭氛围。本建筑的设计通过对世界各地成功的设计方案稍加改变而形成，最著名的案例是由建筑师 Niels Torp 和 Potsdamer Platz 设计的位于伦敦的英国航空公司总部大楼。本建筑的方案就是围绕着建筑师 Renzo Piano 设计的内部中央大街发展而来的。中央庭院是带有水景的摩洛哥式花园，与街景直接联系。另外两个庭院是通往建筑的主要通道，通道通过分层放置的玻璃加强自然光线的射入，从而在高悬挑顶层产生浮动效果。

　　在设计阿曼的建筑时，朝向和阴影极其重要，因为这对于太阳能的利用有重要的影响。建筑外立面的设计将通过几个关键性的设计减小对环境的影响，包括建筑外立面设计特征、改良机械和电气设计的标准以及改善建筑的节水功能。在建筑的初步设计阶段，上述这些是由我们的可持续发展设计团队完成的。北立面和南立面通过悬窗遮阳，东立面和西立面在清晨和下午时通过悬窗使低角度阳光射入。建筑的朝向是北偏西 28°（接近西—北—西），落地玻璃大窗（从地板到天花板）被设在建筑的南立面和北立面。遮阳效果通过设置铝制网屏实现，复制出一种传统伊斯兰式的视觉上的美感。东面和西面的窗户镶嵌在墙壁之中，水平的悬岩将它们连接起来，重要的是，这些窗户和南面与北面的窗户相比尺寸上有所缩小，从而使太阳的照射强度和辐射温度大大降低。建筑的立面旨在使用现代的方式重新定义传统内涵文化。

ALCATEL HEAD OFFICE
阿尔卡特总部

Architects: Frederico Valsassina Arquitectos
Location: Cascais, Portugal
Area: 11,051.81 m^2
Photographer: FG+SG -Fotografia de Arquitectura

设计机构：Frederico Valsassina Arquitectos
项目地点：葡萄牙卡斯凯什市
面积：11 051.81 平方米
摄影：FG+SG –Fotografia de Arquitectura

The building appears as an isolated block, morphologically linked to the ones around it, with a distinctively contemporary architectural language.

Using a volumetrical speech similar to the adjoining volumes, the new building is carved in order to soften and project its image to the outside. The form is enhanced as a projectasset, contributing to the intrinsic dynamic that is intended for the global proposal.

The two levels give place to one, through the "folding up" of the entrance level. The new volume releases itself from the ground, as a business card for anyone who enters. Suspended over the void, it directs people to the entrance, which is made to the north over the void that gives access to the parking lot.

The rotation of the upper volume translates, on the base level, a negative that directly exposes the parking lot, allowing construction to mix with a green protection area. The building is intimately connected with its natural surroundings, taking advantage of it as a barrier against external agents.

In terms of materials, the solution is as restrained as possible to ensure the sobriety of the whole. Therefore, a limited range of materials is introduced in order to emphasize the formal clearance required: white spread surfaces and screen-printed glass are assumed as predominant materials.

Elevation 1 立面图 1

Elevation 2 立面图 2

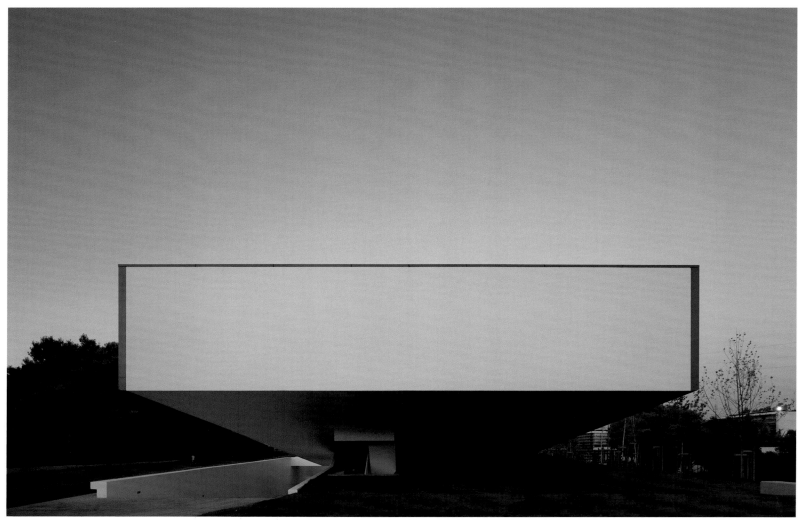

Elevation 3 立面图 3

本建筑看起来像一个孤立的块，其使用独特的现代建筑语言与周围的事物在形态上连接起来。

建筑采用与周围建筑体量相一致的设计。新建筑被精雕细刻，其形象与周围建筑相比更加柔和。建筑的形式作为一种项目资产被加强，为了全球化的目的加强了其内在动感。

通过入口处的"折叠"设计，建筑的两层空间变为一层。新的建筑体量从地面上释放，对于进入其中的人，就像一张商业名片一样，悬空的入口指引人们进入。入口面向北部，使人们方便地出入停车场。

建筑的上层旋转：在基础层，反方向的旋转使停车场显露出来，这将使建筑与被绿色植被覆盖的地面浑然一体。建筑与自然环境亲密连接，并形成一道屏障。

在材料方面，建筑方案尽可能减少材料的使用，以使整体的成本得到控制。因此，有限的材料清晰地表达了形式上的要求。白色表面和印花玻璃是建筑的主要材料。

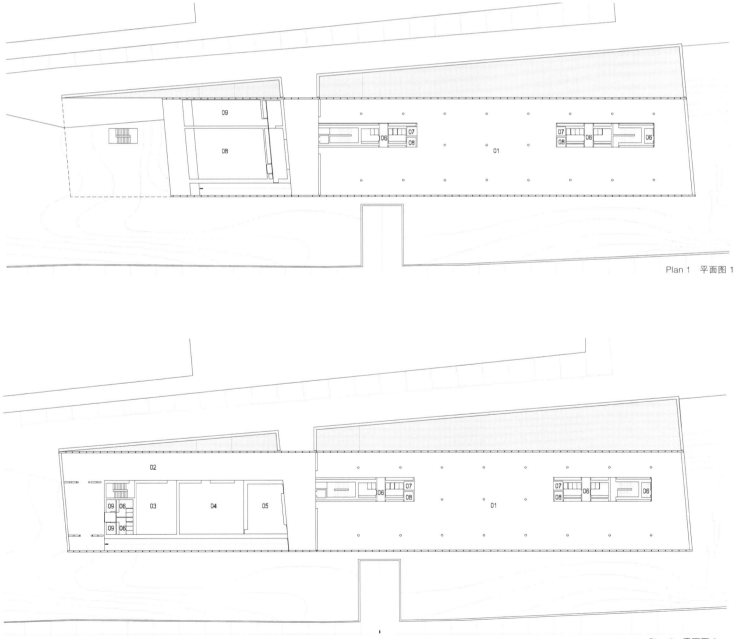

Plan 1　平面图 1

Plan 2　平面图 2

办公建筑   OFFICE BUILDING

ROCHAVERA CORPORATE TOWERS
Rochavera 公司大厦

Architects: Aflalo & Gasperini Arquitetos
Location: São Paulo, Brazil
Area: 33,515m²
Photographer: Daniel Ducci, Nelson Kohn

设计机构：Aflalo & Gasperini 建筑师事务所
项目地点：巴西桑托斯市
面积：33 515 平方米
摄影：Daniel Ducci, Nelson Kohn

Asymmetric and juxtaposed volumes, implementation in non-orthogonal angles, high technology, exuberant landscaping and a philosophy focused on sustainability are the characteristics of the Rochavera Corporate Towers, a Tishman Speyer development by Aflalo & Gasperini Arquitetos.

The project is characterized for a set of office buildings implanted in a site with three fronts: Chucri Zaidan Ave., United Nations Ave. and Parking Street, where it occupies an almost 34,000 m² lot. This situation allows an excellent accessibility to the site, besides the great exposition to the most varied sightseeing spots in the neighborhood, marking definitively its presence in the region.

The project program consists of four buildings, totaling 228,000 m² of built area, which had been implanted around a central square by which the access to the lobby of each set is made, either by pedestrians or vehicles. All inclinations converge to this central square as the implantation allowed to the creation of semi-public urban spaces around all complex, guaranteeing a total integration of the complex with its surrounds.

The towers are distinguished by the slope of nine degrees of the facades, which resulted in the projection of 12 m, creating an elegant and harmonious aesthetic effect. To the architect, the purpose was "using various architectural language, creating a harmonious dialogue between them and escape from the obvious". In addition to the formal intention, the tilted design of the towers has an economic reason: the top floors, always more valued due to the view and noise reduction, are still more profitable with the increase in their dimensions.

The first phase, composed by two of the four towers, with 19 stories in equal and symmetric. Its ground floor is composed by two stores, besides the lobby and the communication stations. The second phase is composed by another two towers, the lower one is 8 stories high and the higher one is 33 stories high. A parking building of five levels is also part of this phase.

Their external facades combine precast elements of concrete covered by polished granite plates, interspersed by a structure of aluminum and covered in glass. A third of the buildings will have the face covered by a curtain of aluminum and glass.

The buildings were designed according to the standards of the Green Building Council, North-American entity conferrthe sustainability LEED GOLD (Leadership in Energy and Environmental Design) certification. As long as Rochavera is considered a sustainable building its characterization is—among other factors— in technological components coupled to the construction, such as water reuse systems, the waste management and energy production as well as in the adoption of architectonic resources given during the phase of the project elaboration.

In summary, Rochavera is an undertaking that is outstanding not only for its social environmental responsibility but, mainly due to its simple architecture, but of great visual impact and thus raising the questions: what would be the ideal city look like? Simple and impressive?

不对称的并列体量、非正交角度、高科技、生机勃勃的景观以及可持续发展的哲学思想是 Rochavera 公司大厦的特点，它是由 Aflalo & Gasperini 建筑师事务设计、Tishman Speyer 房地产开发公司开发的一个项目。

本项目的特点是地块上一系列办公建筑面对着三条主街：Chucri Zaidan 大街、United Nations 大街和 Parking 大街，项目占地 34 000 平方米。这种设计可以使场地更容易进入，更好地展示周围的风光，显现其在该地区的标志地位。

项目由四座建筑组成，建筑面积共有 228 000 平方米。建筑围绕一个绿植覆盖的中心广场建造，广场可供行人或者车辆使用，通向每一个大厅的通道。所有地面倾向会聚到中央广场，允许设计师创造一个围绕所有建筑体量的半开放的城市空间，以保证建筑体量和周围环境的完全融合。

塔楼的外立面具有 9°的倾角，可以形成 12 米的投影，创造一种优雅、和谐的美学效应。建筑的目的是"利用各种建筑语言，为建筑创造一种和谐对话，从而脱离平淡"。除了形式上的意图以外，塔楼的倾斜设计还具有经济上的原因：顶层由于具有极佳的视觉效果，而且降低了噪声，具有较高的价值，因此增加顶层的面积能够获得较高的收益。

建设的第一阶段是四座塔楼中的两座，塔楼有 19 层，都是对称的。建筑的一层包括两个商店、大厅和通信站。第二阶段是另外两座塔楼，较低的塔楼有 8 层，较高的塔楼有 33 层。一个 5 层的停车大楼的建设也在这个阶段进行。

建筑的外墙由抛光花岗岩板覆盖的预制混凝土组成，铝制结构稍加点缀，并覆盖有玻璃。三分之一的建筑被铝框和玻璃幕墙覆盖。

建筑根据绿色建筑委员会的标准设计，并通过了北美可持续发展体系 LEED（能源与环境设计领域的权威）金牌体系认证。Rochavera 公司大厦被认定为可持续建筑，主要是由于它将各种节能技术运用到建筑中：例如水再生利用系统、废物管理系统、能源再生系统以及项目经营阶段的建筑能源供给系统。

总之，Rochavera 公司大厦是一个杰出的项目，不仅仅因为它杰出的社会环境责任，而且由于其建筑形式的简单。巨大的视觉冲击让人们不禁提出了这样的问题：城市看起来究竟像什么？简单还是令人印象深刻？

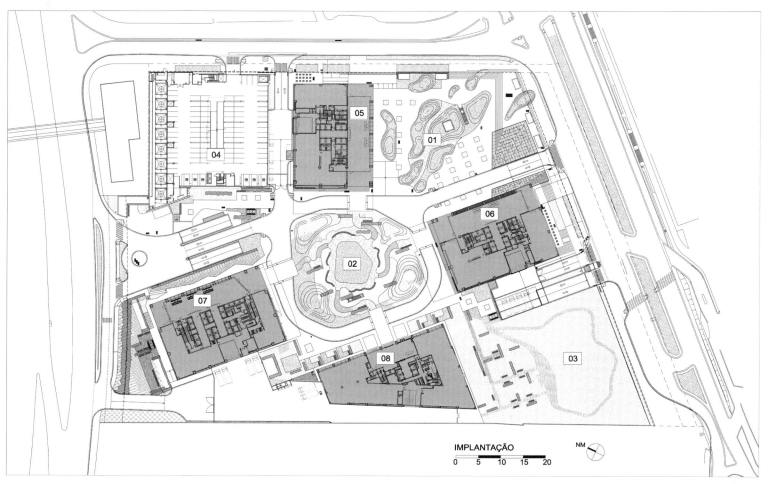

IMPLANTAÇÃO

0 5 10 15 20

NM

Plan 平面图

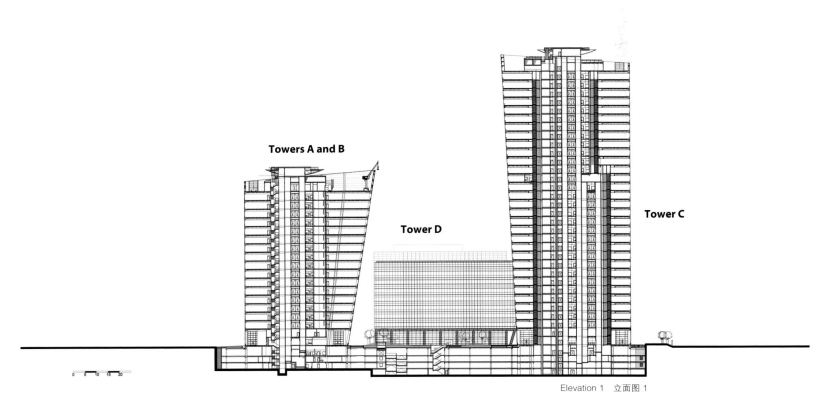

Towers A and B

Tower D

Tower C

0 5 10 15 20

Elevation 1　立面图 1

CORTE LONGITUDINAL

0   5   10   15   20

Elevation 2   立面图 2

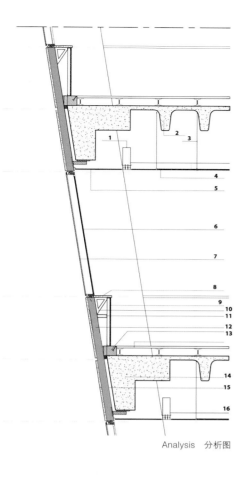

Analysis　分析图

NANJING CHANGFA CENTER
南京长发中心

Architects: WSP ARCHITECTS
Location: Nanjing, China
Area : 140,000 m²
Floor area : 19,160 m²
Photographer:Shu He, Yao Li

设计机构：维思平建筑设计
项目地点：中国南京市
用地面积：140 000 平方米
面积：19 160 平方米
摄影：舒赫、姚力

Nanjing Changfa Center is located in the bustling area of Nanjing City — Daxinggong area, and it is the first-class land used for business and office, with convenient communication and complete auxiliary facilities. The land for the project is located at CBD of Nanjing City, at the east side of the high-rise buildings. Xuanwu Lake — President Building — Confucius Temple form a row of sequence, setting off each other. Nanjing Changfa Center joins in the sequence by letting its axis pass through the middle of the twin office tower. Designer hopes that Nanjing Changfa Center can integrate with the city authentically and become an indispensable part of the city life.

Nanjing Changfa Center is composed of two 150 m high office twin towers, and two 135 m high tower type apartments in the south. The business is concentrated around the sinking square at the foot of the office twin tower in the south and under the huge grass slope which is connected with the city at the foot of the twin tower apartment in the south. Nanjing Changfa Center adopts "low-tech high-efficient" design strategy, i.e. to create highly comfortable and highly efficient environment with simple energy saving and low cost techniques.

The core in core structure system is adopted in the design of Nanjing Changfa Center. The rectangle grid composed by close columns and beams of the outer core can be detected clearly from the facade. The "dual surfaces" structure is adopted by the design for facade: the inner layer is French windows that can be opened and plain frame columns and beams, and the outer layer is large area perforated aluminum panel curtain wall. The inner layer and the outer layer are connected by steel frame. The glass windows that can be opened make the ventilation be realized by opening the windows in high-rise buildings. The perforated aluminum panel can filter the "high building wind" that impacts the building in horizontal direction, and shield 40% excessive sunlight.

The double-height concept is introduced into the interior space of the office and residential towers, with the height of 5.4 meters and 4.95 meters individually, to effectively meet the requirements that the expanding enterprises and families can re-divide the indoor space vertically. At the same time, the rectangular office and residence have high efficient plane and variable and flexible usable space.

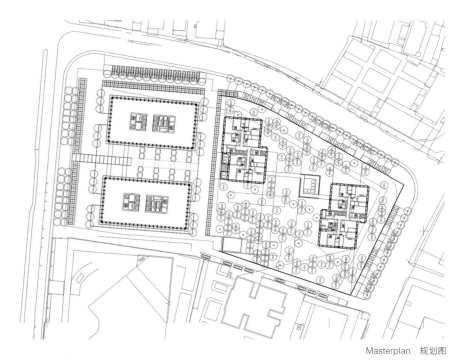

Masterplan 规划图

南京长发中心位于南京市繁华地段——大行宫地段，是南京市一类商业办公用地，交通极为便利，周边公用配套设施齐全。项目用地地处 CBD 超高层区的东端，与玄武湖—总统府—夫子庙形成遥相呼应的序列。南京长发中心以轴线从办公双塔之间居中穿过的方式加入这个序列当中。设计师希望南京长发中心能够真正地与城市融为一体，成为这个城市生活中不可或缺的一部分。

南京长发中心由两栋高 150 米的办公双塔以及南侧两栋 135 米高的塔式公寓所组成。集中商业分别设置于北部办公双塔下的下沉式广场周边以及南部双塔公寓下与城市相衔接的巨大草坡之下。南京长发中心所采用的设计策略之一是"低技高效"，即通过简单的节能材料和低成本的技术营造高舒适度、高效能的环境。

南京长发中心的设计采用钢筋混凝土筒中筒结构体系，外筒间隔较小的横梁和柱构成的矩形网格在立面上被清晰地表达出来。外立面设计采用了独特的"双层表皮"构造，内层是可开启的落地玻璃窗和朴素的框架梁柱，外层是大面积的穿孔铝板幕墙。内外层之间以钢构架相连接。可开启的玻璃窗使超高层也可以通过开窗实现空气流通；穿孔铝板可以帮助过滤横向冲击大厦的"高楼风"，并屏蔽 40% 的多余阳光。

写字楼和公寓的室内空间均采用了双层高度设计，层高分别为 5.4 米和 4.95 米，以便成长的企业和家庭在必要时在竖向上重新划分室内空间。同时，矩形的办公建筑和住宅具有高效率的平面和灵活可变的使用空间。

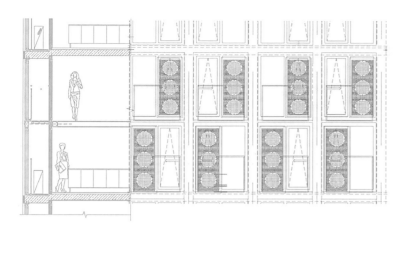

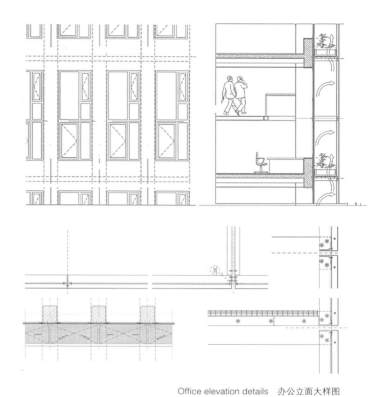

Apartments elevation details 公寓立面大样图

Office elevation details 办公立面大样图

INFINITY TOWER
无限塔

Architects: Kohn Pedersen Fox Associates / Aflalo & Gasperini Arquitetos
Location: Sao Paulo, Brazil
Area: 75,700 m²
Photographer: Daniel Ducci

设计机构：Kohn Pedersen Fox Associates / Aflalo & Gasperini Arquitetos
项目地点：巴西圣保罗市
面积：75 700 平方米
摄影：Daniel Ducci

Infinity Tower is an office tower located in the center of São Paulo, Brazil. Its immediate proximity to the intersection of Faria Lima and Juscelino Kubitschek Boulevard, the heart of São Paulo's Financial District, strategically positioned as one of the city's only Class A office buildings.

The 118 m tall tower elegantly extends skyward from a plaza level reflecting pool and responds simultaneously to both its unique urban context and the city's zoning criteria. The curving reflecting pool gently defines the full height glass enclosed lobby while 18 office floors, each approximate 2,000 m², rise above. The building gracefully wraps each level with two sweeping curves of a glass curtain wall system that is highlighted by exterior 'brise soleil' fins to control sunlight and glare. Private balconies extend from each end of both curves and afford dramatic views of the city's older Avenida Paulista center to the north and the emerging new development to the south. The top of the building is sharply sculpted and punctuated by a rooftop heliport to provide direct executive access to each office level below.

Ground level access is readily accessible to pedestrians and vehicles alike. A series of reflecting pools, shaded walkways, and landscaped green areas offers a quiet transition between the City and Tower. A canopy covered porte-cochere is located directly adjacent to the Lobby's main entry. The nautical motif of the tower also defines the Lobby which is surrounded by water and accessed across two bridges. An 11m high curved glass wall continues the sweeping gestures of the tower above and the material palette features the rich finish of French Limestone for the floor and dark Brazilian wood for the core enclosure.

The project is complete and on track to achieve LEED Gold Status.

　　无限塔是一座位于巴西圣保罗市中心的办公大楼，靠近法里亚利马和儒塞利诺·库比契克大道的交叉路口，贴近圣保罗金融区的心脏地带，战略定位为该市的甲级办公大楼之一。

　　这座118米高的建筑从底层优雅地向上延伸，反射出广场上水池的同时，呈现出独特的城市环境和城市的分区标准。有弧度曲线的水池衬托着拥有落地玻璃的大厅，大厅上面有18层楼的办公室，每一层都接近2 000平方米。建筑用两个曲线形玻璃幕墙系统优雅地围绕每一层，以突出外部的"鳞状光影"，控制光线和强光照射。私家阳台从每个弧线的末端伸出，坐拥北部城市旧街保利斯塔中心的美景，同时也能看到南边正在兴起的新发展区。建筑的顶部被大幅雕刻，又被顶层的直升飞机场打断，提供了直接通往楼下每一层办公室的通道。

　　行人和车辆可以很方便地到达地面入口。一系列水池、林荫道、绿化带为城市和大楼之间提供了幽静的联系纽带。顶棚覆盖在车辆通道之上，毗邻大厅的主入口。大楼的航海主题也更好地定义了经由两座桥方可到达的绿水环绕的大厅。11米高的曲线形玻璃幕墙继续演绎着大楼的霸气，在材料上，地板运用了法国石灰石，核心围墙采用了巴西木材。

　　此项目已经完工，正在申请LEED绿色建筑的金牌认证。

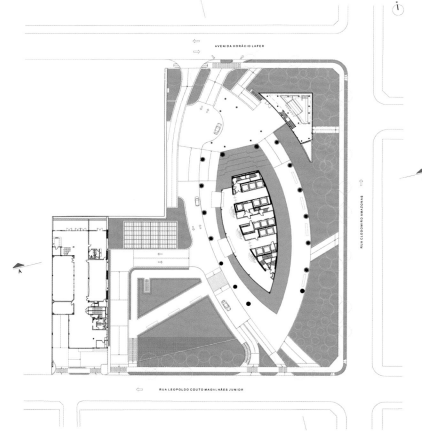

Ground Floor Plan　底层平面图

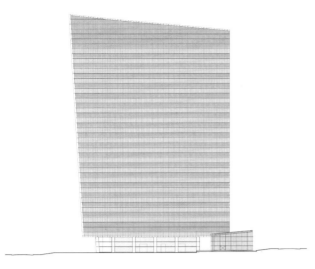

East Elevation　东立面图

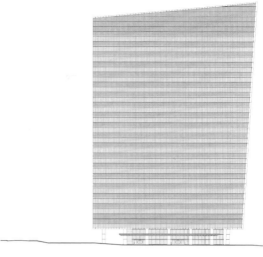

West Elevation　西立面图

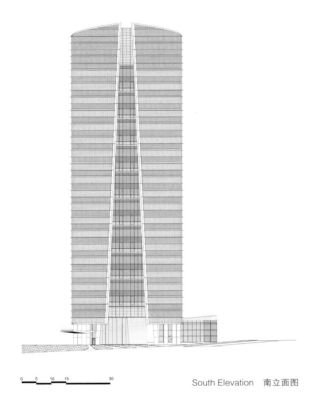

South Elevation　南立面图

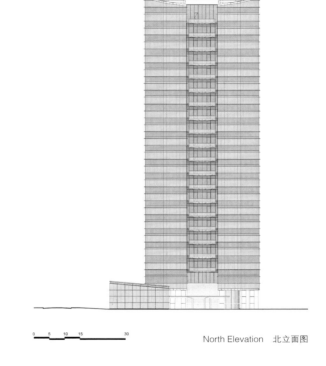

North Elevation　北立面图

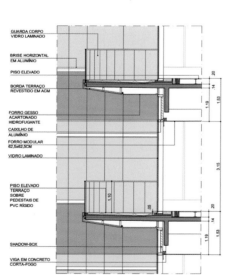

Typical Section 01-North Facade　标准剖面图 01- 北立面

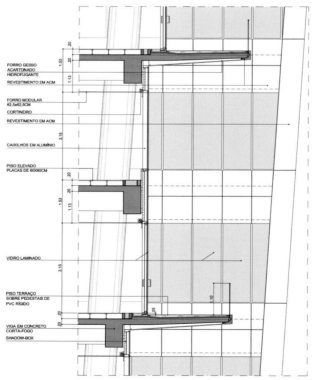

Typical Section 01-South Facade　标准剖面图 02- 南立面

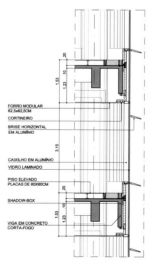

Typical Section 03-East / West Facade
标准剖面图 03- 东西立面

Facade details　立面细节图

MENARA KARYA
摩洛哥风情城

Architects: Arquitectonica
Partners-in-Charge of Design: Bernardo Fort-Brescia, FAIA and Laurinda
Spear, FAIA, ASLA, LEED AP
Interior Designer (Public Area): Arquitectonica
Location: Jakarta, Indonesia
Area: 52,200 m²
Client: PT Karyadeka Pancamurni
Photographer: Ray Sugiharto

设计机构：Arquitectonica
协助设计：Bernardo Fort Brescia，美国建筑师协会等
内部设计（公共区域）：Arquitectonica
项目地点：印度尼西亚雅加达市
面积：52 200 平方米
客户：PT Karyadeka Pancamurni
摄影：Ray Sugiharto

Menara Karya is a 26-storey, 52,200 m² commercial office development, located in Jakarta's prestigious "Golden Triangle", the heart of Indonesia's commercial and financial district. It also forms the next phase to Menara Kadin, already a premier address in the city's business community.

The return of signature architecture to Jakarta's skyline signals the city's and the country's return from the prolonged doldrums that followed the onset of the Asian economic crisis of the late 1990s. In a desire to stamp an identity on the recovering market, developers are deviating from tried and tested patterns for local commercial developments that are designed to stand out among the crowd of newcomers.

Menara Karya signals a new direction for the client's business aspirations. It is a progression from mainstream corporate to a more expressive, individual aesthetic, and a signal for more powerful architectural identities in order to lead in the region's competitive markets.

The exterior of the building has been sculpted in an angular, chiseled form, much like an abstract diamond. This creates a powerful, unique profile for the tower which sets it apart from the uniformity of rectilinear forms on Jakarta's skyline.

The client's briefly demanded efficiency and flexibility for the international tenancy market, Grade A design and technical standards, in addition to projecting iconic imagery. The result was simple rectilinear floor plates ranging from 1,300 to 1,400 m² / 13,993~15,069 sq. ft. in size, with 13.5 m / 44.3 ft. clear span from exterior to core. Column free floor and corners, and floor to ceiling glazing allow for bright and expansive tenancies.

The pure crystalline form rises directly from ground, allowing the lobby and office floors to be unhindered by any above-grade parking podia that is endemic to many local high-rise commercial developments. Such podia may be cost effective, but compromise the on-grade open space and quality of public space, which is why it was avoided for this project. Parking is efficiently contained in 3 basement levels, and permits the pure building form to sit elegantly in spacious and lushly landscaped grounds.

With the dramatic facades continuing unbroken to the ground, the entrance is called out by a planar wing-shaped canopy of fritted glass, clipped to the building's base and corner. The randomly balanced forms extend to the interior, where the tall main lobby interiors take a cue from the sculpted exterior: 2-storey high angular facets are repeated on the lobby wall where they are expressed in a bold but warm travertine finish. Double height portals break this inner facade, which define access to lift lobbies and tenancies. The main lobby ceiling is also composed of simple angular plates which align with the walls. Four tones of granite are used on the lobby floor in a parallel barcode pattern, which maintains it's continuity from the exterior paving through the lobby, lift cab and to public tenant spaces. Interiors evoke a warm corporate design, while random complementary surfaces further reinforce the distinct identity established by the exciting exteriors.

摩洛哥风情城是一座 26 层、建筑面积达 52 200 平方米的商业建筑,位于雅加达著名的"金三角"地区、印度尼西亚的商业和金融中心。它也是摩洛哥印尼工商会馆的下一阶段建设内容,是城市商业社区的首要建筑。

标志建筑重现于雅加达的天际线标志着城市和国家已经从 20 世纪 90 年代亚洲金融危机经济长期低靡的萧条中恢复过来。为了给逐步恢复的市场树立一个丰碑,开发商们放弃了当地商业模式发展的尝试和摸索阶段,目的是在后来者中脱颖而出。

摩洛哥风情城标志着客户商业精神的新方向。它是从主流企业到更富于表现力的个人美学的发展,是更强大的建筑特征的标志,用以领导当地的竞争市场。

建筑的外观被设计为有棱有角且轮廓分明的形式,就像一颗抽象的钻石。这将为塔楼创造一个强大的、独特的外立面,在雅加达的天际线上创造出与其他的线性建筑相区别的新颖建筑。

客户要求这个国际化租赁市场应具有高效性和灵活性,建筑需要具备甲级的设计和技术标准以及标志性的外形。建筑使用了简明的波纹板,面积为 1 300~1 400 平方米,从外部到核心筒跨度为 13.5 米。楼层的底板和转角没有柱子,从地板到天花板的空间允许租户方便地利用。

纯水晶的形式从地面直接上升,使大厅和办公层不受任何交易受理台的影响,这是许多商业建筑普遍面临的问题。这样的设计十分有效,而且为内部空间让步,保证了公共空间的质量,这就是为什么这个项目能避免以往问题发生的原因。

引人注目的外立面具有连续感,建筑的入口被二维的翼形烧结玻璃覆盖,一直到建筑的基础和转角。建筑外部的随机平衡形式延伸到室内,内部的建筑大厅采用了建筑外表面的形式:两层高的有棱角的表面在大厅的内墙面上重复,覆盖在粗犷但是温暖的石灰石材质的墙壁上。双层高的大门打破了建筑的内表面,提供了通往电梯和租屋的通道。主体大厅的天花板由简单而有棱角的平板组成,大厅的地板使用了四种不同的花岗岩,这些花岗岩平行叠放,保持了从外部装饰到大厅、电梯出口和公共租户空间的连续性。内表面呈现了温暖的公司设计,而随机互补的表面加强了建筑外表面的特征。

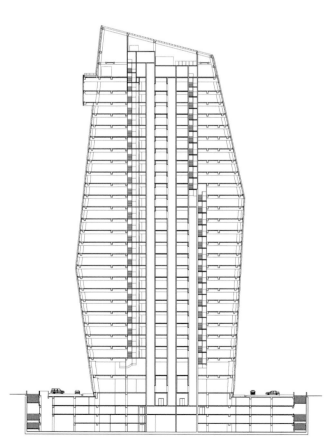

SECTION

1 5   20m
0 2   10

Section 1 剖面图 1

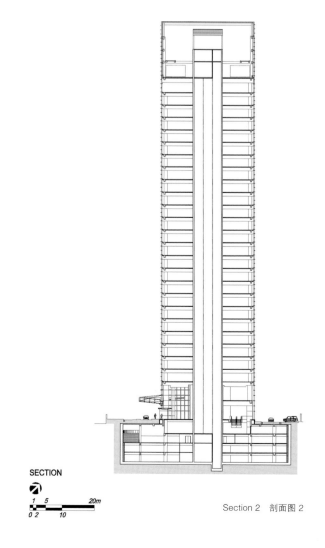

SECTION

1 5   20m
0 2   10

Section 2 剖面图 2

GUANGSHENG INTERNATIONAL TOWER
广晟国际大厦

Architects: Guangzhou Hanhua Architects Engineers Co., Ltd.
Location: Zhujiang New Town, Guangzhou, China
Area: 155,635 m²

设计机构：广州瀚华建筑设计有限公司
项目地点：中国广州市珠江新城
面积：155 635 平方米

Guangsheng International Tower regarded as intelligent super grade A office building will be one of the landmarks in Guangzhou Zhujiang New Town CBD in the future. Total height of the building is 312 m and the height from roof top to the ground is 350 m. Vertical strips are used as the composition theme for the facade and gradation is enriched through terrace on the top and materials such as stone, glass and metal. The building is tall and dynamic, dignified and keen, reflecting the noble architectural style of neoclassicism.

The design follows the concept of environmental protection, and properly controls curtain wall area. Pollution caused by jointing in old installation technology is avoided through advanced structure technology. LOW-E double glazing unit, light-colored stone and materials with low heating co-efficient are adopted for the external wall, which can effectively reduce energy consumption of the building.

定位为智能化超甲级写字楼的广晟国际大厦，是广州市珠江新城 CBD 核心未来的标志性建筑之一。建筑总高度为 312 米，至屋面标志杆顶部达到 350 米。立面以竖向线条为构图主题，并通过顶部退台变化以及石材、玻璃、金属等材质的对比来丰富层次。建筑造型挺拔而富有韵律感，自稳重中散发新锐气息，诠释高贵的新古典主义建筑风格。

设计遵循环保理念，适当控制幕墙面积，并通过先进的构造技术来避免旧式安装工艺中由填缝工序带来的污染。外墙采用 LOW-E 双层中空玻璃、浅色石材和低采暖系数材料，有效降低了建筑能耗。

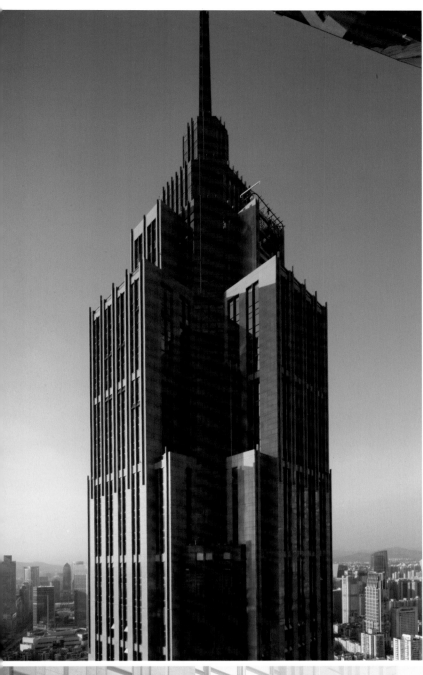

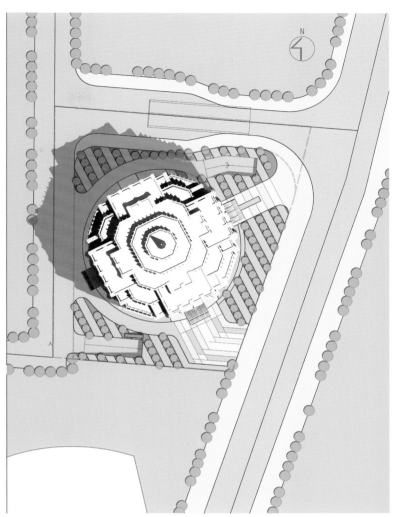

Plan 平面图

ADMINISTRATION CENTER OF NANXUN DISTRICT, HUZHOU

湖州市南浔区行政中心

Architects: China United Zhujing Architecture Design Co., Ltd.
Location: Huzhou, China
Area: 23,094.8 m²

设计机构：中联筑境建筑设计有限公司
项目地点：中国浙江湖州市
面积：23 094.8 平方米

The ancient town of Nanxun has a long history and cultural deposit. It is the key point in the design to perfectly combine culture of Jiangnan region with the solemn style of administrative buildings.

The so-called "New Chinese Building" still rests on the level of mannerism. Actually, traditional architectural space pays more attention to the association with building function and aesthetics; moreover, there is no concept of individual building in Chinese traditional architecture which values the establishment of group building spatial system. At the same time, one spatial system is contained in a larger system and it is actually a "self-similarity" structure from individual to group; the design concept and connotation of single building and group building actually come from the same source.

Apart from the courtyard and surrounding space, import of a series of modern materials like glass and steel also provides an exuberant sense of the times for the building. Connotation of Chinese gardens is unfolded like a picture scroll between "similarity and dissimilarity".

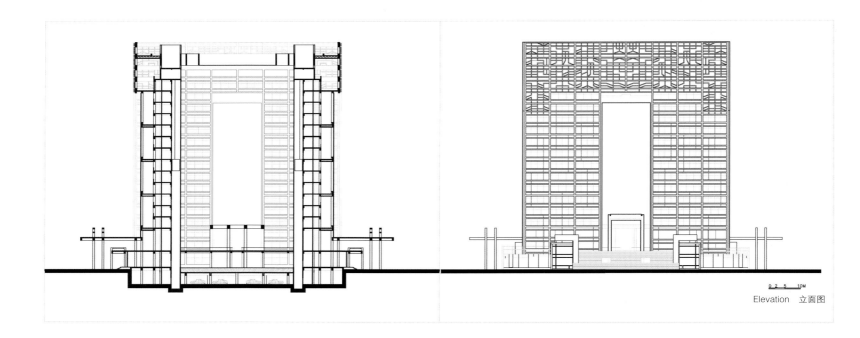

Elevation　立面图

　　南浔古镇历史久远，文化底蕴深厚。如何使南浔江南水乡文化与行政建筑开放庄重的气质完美结合，是设计中的核心问题。

　　当下所谓的"新中式建筑"，基本仍然停留在手法主义的层面上。实际上，传统建筑空间更加注重一种与建筑功能、审美相联系的意境表达，而且，中国传统建筑基本上不存在单体建筑的概念，侧重的是群体建筑空间体系的建立。同时一个空间体系又包含在一个更大的体系之中，由单体到群体其实是一个"自相似性"的结构，建筑单体与群体的设计概念和寓意其实是同根同源的。

　　院落萦回，空间环绕，玻璃、钢构等一系列现代材料的引入又使建筑平添了一种生机盎然的时代感。中国园林的意境就在"似与不似之间"如画卷般展开……

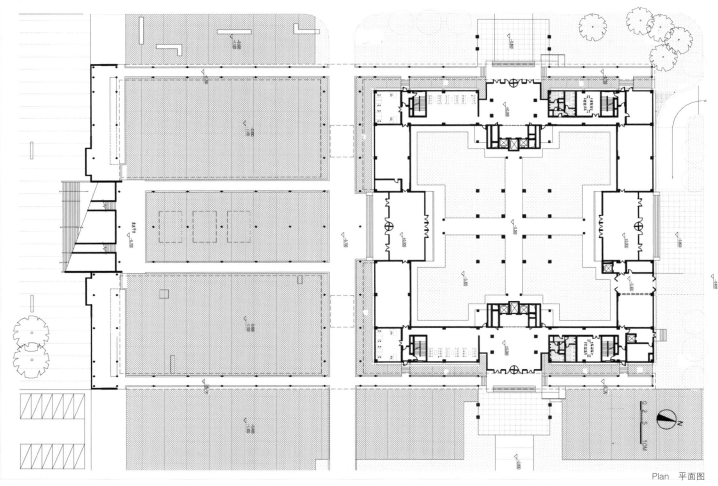

Plan 平面图

HEADQUARTERS BASE OF SHENZHEN CADRE GROUP CENTER

深圳凯达尔集团中心总部基地

Architects: Shenzhen Urban Space Architectural Design Consulting Co.,LTD
Location: Shenzhen, China
Area: 67,900 m²

设计机构：深圳市都市空间建筑设计顾问有限公司
项目地点：中国深圳市
面积：67 900 平方米

When the design company accepted this project from the proprietor in early 2007, it was initially positioned as a R&D and production base. Additionally, considering that the proprietor is a high-tech enterprise engaged in the R&D of intelligent transportation, it has different requirements with common office buildings or R&D bass, it is mainly reflected in three aspects that are high standard, high starting point and intelligentization. Therefore, both the inner space and the external appearance shall meet actual demand as well as reflect the corporate image.

Considering the terrain elevation difference in building group plan, it adopted a semi-buried ecological garage design to create a unique sloping external environment space, thus reducing the amount of earthwork. After the completion of the construction ,the site landscape was recovered as before.

To create an upright and spacious office room under the limitation of standard layer contour area, the high-rise core-tube is designed to have a bias effect, create a standard layer office plane with a large interior span structure. Meanwhile, considering the structural balance in the west exterior wall with bias core-tube, multiple upright columns are applied there, thus achieving the effect of completeness and unification from inner space to external structure, to create an image of dissymmetrical with the sense of balance.

There is a multiple function hall in the podium of the main building, while a glass inner sky bridge nearby links the main tower and secondary tower, which becomes the visual focus among the group towers, and releases the pressure from the huge mass of the twin towers in the outer space.

The external wall adopts stone curtain wall system of various modulus collocations, thus making the concise building mass full of rich and changeable layer.

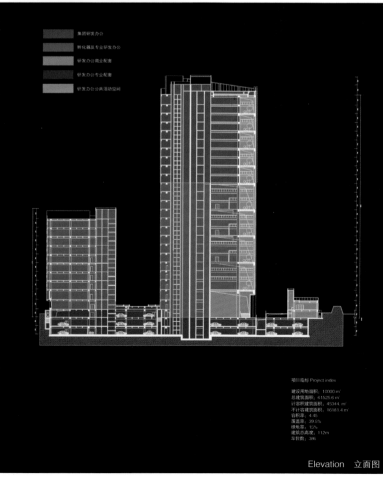

项目指标 Project index

建设用地面积: 10000 ㎡
总建筑面积: 61525.6 ㎡
计容积建筑面积: 45344 ㎡
不计容建筑面积: 16181.4 ㎡
容积率: 4.45
覆盖率: 39.5%
绿地率: 15%
建筑总高度: 112m
车位数: 386

Elevation 立面图

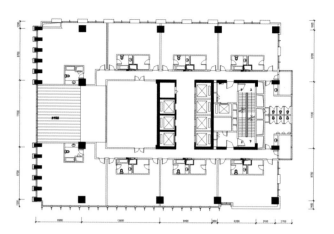

1栋4、7、10、13F平面图（带空中花园）7、10、13F平面图（带会议中心）

Plan 1　平面图 1

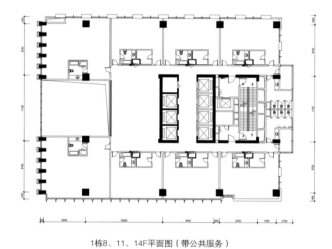

1栋8、11、14F平面图（带公共服务）

Plan 2　平面图 2

2007年初，设计公司接受业主委托，对项目进行了多轮的布局探讨和定位修改，最初定位为研发生产基地。业主为从事智能交通研发的高科技企业，对项目的要求不等同于一般的办公楼或者研发基地，着重体现在高标准、高起点和智能化三方面，希望在内部空间和建筑外观方面均要满足实际需求和彰显企业形象。

建筑群体规划结合现状地形高差，营造独特的坡地外部环境空间，采用覆土半埋式的生态地面车库设计，减少了基地的土方开挖量。施工完成后，对基地原有山地地形进行了最大限度的还原。

虽然由于用地条件的限制，高层办公部分的标准层轮廓面积较小，但是为了达到标准层办公空间较大进深的使用要求，将高层内部的电梯井道核心筒偏置东侧，西侧留出内部跨度较大的方正的标准层办公平面。同时考虑到偏置核心筒后的结构平衡，在西侧外墙安排竖向密柱，既减小了西侧外墙的开窗面积，又创造出整体建筑的不同寻常的不对称均衡感和挺拔感，追求建筑从内到外、从内部结构到外部形象的真实统一。

建筑裙房设有多功能报告厅，并在附近设置空中室内连廊联系主楼与副楼。空中连廊也成为建筑群中的视觉焦点，缓解并释放了两侧较高的建筑体量所带来的外部空间压力。外墙采用多种模数搭配的石材幕墙体系，使简洁的建筑体量体现出丰富多变的层次感，彰显出高科技企业总部基地的气质。

TENCENT BUILDING
腾讯大厦

Architects: CCDI
Land area: 5,999.85 m²
Total construction area: 88,180.38 m²
Increased area approved by the building authority (Refuge floor):2,336.69 m²
Building area: 2,390.55 m²
Height: 193.2 m
Location: Shenzhen, China

设计机构：悉地国际
用地面积：5 999.85 平方米
总面积：88 180.38 平方米
核增面积（避难层）：2 336.69 平方米
建筑占地面积：2 390.55 平方米
建筑高度：193.2 米
项目地占：中国深圳市

"Software Product R&D Center of Tencent Technology" is located in the north side of Shennan Boulevard in Gaoxin District, Shenzhen. It occupies a land area of 5,999.85 m² and a total construction area of 85,843.69 m². The building is composed of a tower formed by the office space of 39 floors and the annex of 2 floors; with a height of 174 m, it belongs to the super high-rise buildings, with two refuge floors established on the 15th and 30th floors. Three floors of parking lots which hold 300 stalls as well as equipment rooms are set underground. The entire building is exclusive for Tencent and symbolizes the development of Tencent.

Main characteristics of the project: it is the corporate headquarter and the super high-rise building beside Shennan Road which is the arterial road in Shenzhen. Therefore, our creation is aimed at four key points.

1.Full of modernity.

2.A landmark to highlight individuality of Tencent.

3.A sense of pride for the owner.

4.Full of imagination.

"Concreteness" should be avoided for the corporate image that cannot be too superficial either. Construction means should be "abstracted" by grasping characteristics of the company or its famous products.

Tencent is a famous developer and operator of instant messaging software on internet; it is a young company full of vigor, showing aspirant spirit.

Under such inspiration, advanced computer surface aided design was adopted to create an architectural image that stands towards the cloud like a drawn sword; slightly inward curves on northern and southern facades have ingeniously avoided bloat and presented elegance. The architectural image is equipped with both industrial beauty and architectural beauty.

Tencent is a listed enterprise in IT industry and the corporate spirit is: forging ahead and pursuing excellence; the headquarter building should fully display this spirit.

The building masses of this scheme have different heights, which has formed a rising body, powerful and aspirant.

Vision of Tencent: the most reputable internet enterprise.

Our target: landmark that presents classic beauty.

设计灵感源自流畅线条，建筑师从腾讯 "QQ" 企业形象中提取
出曲线元素，巧妙的将其运用在建筑造型上，使建筑有充
满向上延伸感和动感的腰线。建筑造型富有想像力。

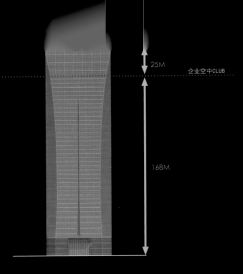

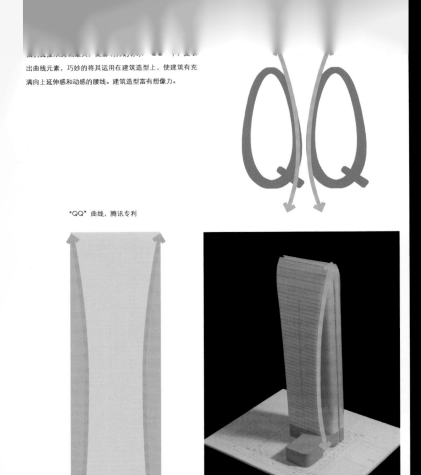

"QQ"曲线，腾讯专利

符合空间流体力学　平剖面都是凸曲线

25M
企业空中CLUB

168M

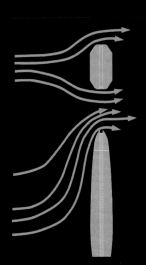

Analysis　分析图

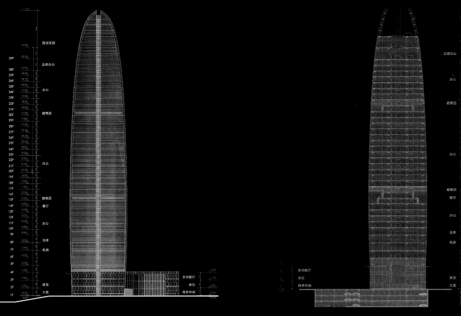

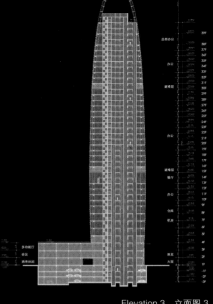

Elevation 1　立面图 1

Elevation 2　立面图 2

Elevation 3　立面图 3

"Proportion" is the keyword of architectural aesthetics and the focus of classic architectural image; it is especially important for super high-rise buildings.

Proportion is redivided via multimass combination technique in this scheme. The main facade has a symmetrical proportion and the side facade is high and steep; the entire building is elegant in an ideal proportion. The magnificent building shape with beautiful proportion will stand the test of time.

The project is located on "the north side of Shennan Boulevard", adjacent to high-rise buildings in the west and east; it is a member of buildings beside the urban artery. In such urban relation, the shape of "four same facades" or four continuously variable facades is inappropriate. The building should have both "main facade" and "side facades", with "side facades" responding to each other and "main facade" displaying characteristics, to form a building with both order and characteristics. Northern and southern facades of buildings in this scheme are in shuttle shapes. On one hand, it can guarantee square and useful architectural plane; on the other hand, it is smooth along the horizontal direction, which can send eastern and western buildings in order; meanwhile, it is tall and straight when viewed from Shennan Boulevard; the height of over 190 m has enhanced such subtle change.

According to practical situations, the following new technologies and new materials are adopted in this project:

1. Mechanical connection is adopted to connect vertical rebar, which can guarantee quality of rebar joints and reduce rebar quantity for joints.

2. HRB400 is preferred as large diameter rebar. Structural design will match up with architecture, and requirements of building functions and creativity should be satisfied as far as possible under safe and reasonable conditions.

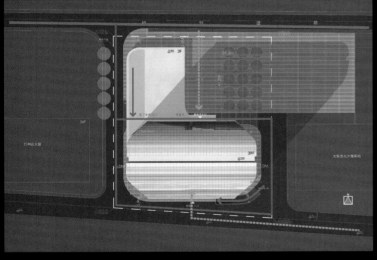

Plan 1　平面图 1

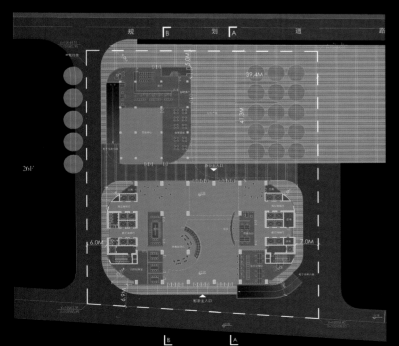

Plan 2　平面图 2

"腾讯科技公司软件产品研发中心"位于深圳市高新区深南大道北侧。项目占地 5 999.85 平方米，总建筑面积为 85 843.69 平方米 。其由 39 层办公空间形成的塔楼及 2 层裙房组成，高 174 米，为超高层建筑，两个避难层设于第 15、30 层。地下设 3 层拥有 300 个车位的停车场及设备用房。整个建筑完全供腾讯公司自用，是腾讯发展的象征。

本项目的主要特点：它是企业总部，是深圳主干道 ——"深南路"边的超高层建筑。因此，我们的创作有三个关键点。

1. 充满时代感。

2. 彰显腾讯公司企业个性的标志性建筑。

3. 令拥有者自豪。

4. 极富想象力。

反映公司形象应避免"具象化"，避免简单肤浅，而应把握公司或其著名产品的特点，以建筑手法"意象化"。

腾讯公司是著名的互联网即时通信软件开发及运营商，年轻充满活力的公司无处不透露着奋发进取的锐气和冲劲。

受此启发，我们运用先进的电脑表面辅助设计，创造出如出鞘利剑般直冲云霄的建筑造型。南北立面微微收分的曲线巧妙地避免了体量的臃肿感，灵动优雅。建筑形体兼具工业美感和建筑美感。

腾讯公司是上市的 IT 行业企业，公司的企业精神是：锐意进取，追求卓越。总部大楼应充分体现这一精神。

本方案建筑体量高低有致，形成强烈的上升气势，取大气有力、锐意进取之势。

腾讯公司的愿景：最受尊敬的互联网企业。

我们的目标：体现经典之美的标志性建筑。

"比例"是建筑美学的关键词，是经典建筑形象的重点，对超高层建筑尤为重要。

本方案以多体量组合手法重新划分呈形体比例，主立面比例匀称，侧立面高耸挺拔，整个建筑比例理想，优雅修长。建筑体形端庄得体，其比例之美，必经得起时间考验。

本项目位于"深南大道北侧"，东西侧各有高层建筑相邻，它是城市干道边建筑行列的一员。在这种城市关系中，"四面同性"或四面连续变化的造型并不合适，应让建筑有"主面"、"侧面"之分，"侧面"彼此对应，"正面"展示特点，形成既有序又有特点的建筑行列。本方案建筑南北立面微微收分呈梭形，一方面这种手法能保证建筑平面方正好用；另一方面，体形水平向平整，利于使东西建筑形成建筑行列秩序；同时，从深南大道上看，形体挺拔锐利，190 多米的高度使这种微妙变化韵味悠长。

根据实际情况，本工程采用以下新技术、新材料：

1. 竖向钢筋驳接采用机械连接，可保证钢筋接头的质量和减少接头的钢筋用量。

2. 大直径钢筋优先选用 HRB400。结构设计将积极配合建筑专业，在安全合理的条件下，尽量满足建筑功能和创造性的要求。

SHANGHAI CENTRAL ENTERPRISE PLAZA

上海中环企业广场

Architects: Tianhua Architecture Design Co., Ltd.
Location: Shanghai, China
Area: 81,536 m²

设计机构：天华建筑设计有限公司
项目地点：中国上海市
面积：81 536 平方米

The project is located in Zhabei District, Shanghai City, with Paper Machinery Co. Ltd. of Shanghai Electric Group to the east, Shanghai Sifang Boiler Factory to the south, Gonghe New Road to the west, and Papermaking Co. Ltd. of Shanghai Electric Group to the north. The single office building of the project faces the south and it is combined with the community layout, with its height less than 100 m and floor height of 5.50 m. The ground floor supports service facilities, the office building lobby is located on the 1st floor in the west side of the building. Two floors are combined together, with a height of 11m and an area of about 500 m². The typical floor area is about 1,500 m², there are 6 elevators and 1 goods elevator with a weight of 1,600 kg, and public lavatory and water heater room are set on the floor. The core tube is arranged on the north side and a small office is designed in the north, while a big office is set in the south; in the future, they are expected to connect each other. Central Enterprise Plaza is planned with the idea of creating good quality. The loft product with a height of 5.5 m has become the benchmark when 5A office building is sold in Central Enterprise. A flexible space of 73-151 m² is the first choice for U-Business and provides an excellent working environment for high-end office; 5A+ intelligent configuration is suitable for various requirements of high-end office; the lobby with a height of 11 m and exclusive elevator for VIP and presidents reflect honor of the enterprise. Besides, transportation system composed of No. 1 metro line, central line and south-north viaduct makes it quite convenient for travelling, which has highlighted superiority of Central Enterprise Plaza again. The compound office space with a height of 5.5 m in Central Enterprise Plaza is also rare among all office properties in Inner Center, which has helped Central Enterprise Plaza achieve the chief position in 5A office loft.

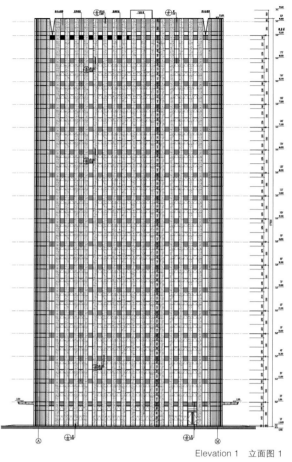

Elevation 1 立面图 1

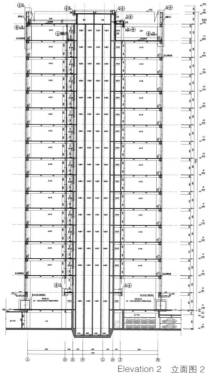

Elevation 2 立面图 2

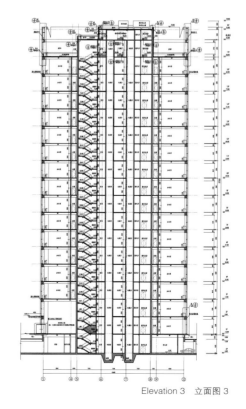

Elevation 3 立面图 3

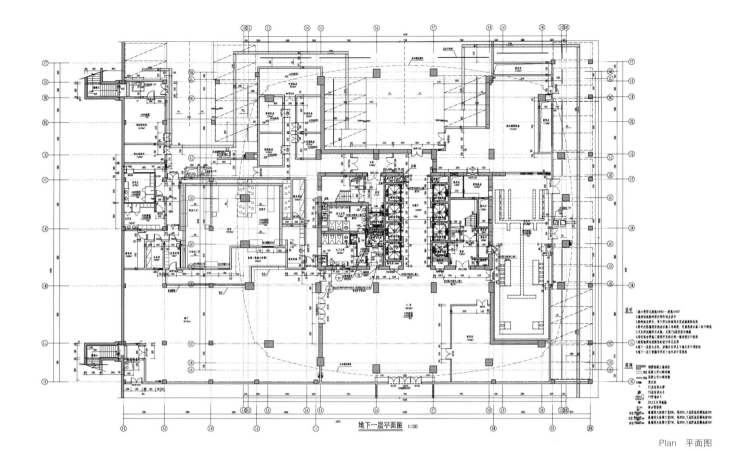

地下一层平面图 1:100

Plan 平面图

该项目位于上海市闸北区，其用地位置东至上海电气集团造纸机械有限公司，南至上海四方锅炉厂，西至共和新路，北至上海电气集团造纸有限公司。本项目中的单幢办公楼正面朝正南方向，和小区布局相结合，高度小于100米，层高5.50米。底层设置服务性设施，办公楼大堂放在一层，位于大楼的西侧。两层的高度挑空，高度为11米，面积在500平方米左右。标准层面积为1 500平方米左右，有6部客梯和1部货梯，电梯载重1.6吨，设公共厕所和开水间。核心筒放在偏北侧，北侧做小空间办公，南侧做大空间办公，并考虑今后可打通。中环企业广场位于大宁国际、市北高新双核之心，未来将以这里为中心，形成一个新的商务CBD。中环企业广场秉承着打造精品的理念来规划。5.5米层高的loft产品成为中环在售5A写字楼的标杆之作。73~151平方米的灵动空间是优化商务之选，为高端办公保驾护航；5A+智能化配置适应各类高端办公需要；11米挑空大堂、VIP总裁专享电梯，彰显企业总部的尊荣内涵。此外，地铁1号线、中环线、南北高架所构成的便利交通体系，更加突出了中环企业广场的优势。中环企业广场所具有的5.5米层高的复式办公空间在内中环的办公物业中是罕见的，这也成就了中环企业广场在内中环5A办公loft中的首席地位。

综合建筑 COMPLEX BUILDING

After the era that the distinction between the residential real estate and commercial real estate is strict,the single real estate development mode could no longer satisfy the residency requirements of people's lives,also could not adapt to the rapid development of the real estate.Creating a new real estate development mode has become a new issue that developers need to facing. Mixed–using building integrates many functions of the city, including commerce, office, residence, hotel, exhibition, catering, conference, entertainment and transportation, and establishes a dynamic relationship of interdependence in various parts,then forming a versatile, high–efficiency building community. As the urban mixed–using building has all functions of modern city, they are often called as "city within a city".

A successful mixed–using building need to complete with four points.First of all, it needs to be creative and correspond with the external shape of the era aesthetics. Secondly, it needs to make good mixed–using planning, scientific and reasonable arrangement of residence, hotel–style apartments, office buildings, shopping malls, hotels, theme parks and other projects. However, these depend on the local urban development process, family income, residents' purchasing power and consumption structure. Thirdly, it needs to reasonably arrange the stream of people, traffic organization, and combination of space, etc. Different industries do not interfere with each other, making complex really play a compact, comprehensive role. Finally, it needs to pay attention to the combination of architecture and ecology, to increase the green area and achieve humanistic ecology. So all kinds of people in the complex could live together with each other harmoniously.

Mixed–using buildings are not only the landmark buildings of city, but also have been the standard of international life style system of each big city's commercial center district.They are the symbol of high quality city life, they can improve the city's overall image quality and value.The complex of mixex–using building determines that it has strong social function which is the engine of regional economy and the main factors that could enhance urban economy and culture. Excellent commercial complex can enhance the flow of the city, improve the quality of life and consumption,and become an important tourist attraction in the city. All these can enhance the attraction of the city.

The development of the modern city is transformed from extensive to intensive direction. Urban mixed–using buildings are more and more emphasizing on the city's openness and integration, focusing on the construction of urban public space, which forms city's public spaces, such as the transportation hub, cultural square.Through the linkage with city, regional development has been integrated.

　　在经历了将住宅地产和商业地产严格区分的时代后，单一的地产开发模式已经越来越不能满足人们的生活居住要求，也不能适应地产的迅速发展，开创一种崭新的地产开发模式就成为地产商们面临的新课题。城市综合体是将城市中的商业、办公、居住、旅店、展览、餐饮、会议、文娱和交通等城市生活功能融为一体，并在各部分间建立一种相互依存、相互助益的能动关系，从而形成一种多功能、高效率的建筑群落。由于城市综合体基本具备了现代城市的全部功能，因而往往被称为"城中之城"。

　　一个成功的城市综合体项目，关键要做好四点。首先是要有独具创意、符合时代审美观的外在形态；其次，做好城市综合体的规划设计，科学合理地安排住宅、酒店式公寓、写字楼、商场、酒店、主题公园等多种项目，而这些都要根据当地城市的发展进程、家庭收入情况、居民的购买力和消费结构来决定；第三，要做好人流的组织、交通的组织、空间的组合等，不同业态之间要做到互相不干扰，使综合体真正发挥紧凑的、综合性的功能。最后，要注重建筑与生态的结合，增加绿化面积，实现人文生态，使综合体中的各种人群和谐相处。

综合体不仅是城市地标性的建筑，而且已经成为各大城市商业中心区的标准化国际通行生活模板体系，是高品质城市生活的标志，能够提升城市价值、城市品质和整体形象。城市综合体的多重复合性决定了其具有很强的社会功能性，是拉动区域经济的引擎和城市经济文化的增级场。优秀的商业综合体能够增强城市的流通力，提高生活和消费品质，也能够成为城市里重要的旅游景点，有利于增强城市吸引力。

　　现代城市的发展是由粗放型向集约型方向转化的，城市综合体越来越多地强调对城市的开放性与融合性，注重城市公共空间的建设，形成城市的公共空间，如交通枢纽、文化广场等，通过与城市、区域的联动发展，实现一体化。

BANCO SANTANDER HEADQUARTERS TOWER—WTORRE PLAZA
桑坦德银行总部大厦——WTORRE 广场

Architects: Arquitectonica
Designers: Bernardo Fort-Brescia and Laurinda Spear
Location:Sao Paulo,Brazil
Area: 411,975 m²
Photographer: Francisco Donadio

设计机构：Arquitectonica 建筑事务所
设计师：Bernardo Fort–Brescia and Laurinda Spear
项目地点：巴西圣保罗市
面积：411 975 平方米
摄影：Francisco Donadio

This mixed-use project involves new construction of a shopping mall and two office towers and the facade and interiors design for an existing structure to be fitted out as an office tower. The new building of totally 105,701 m² area includes a 66,183 m² Iguatemi Shopping Mall, a 22-storey office tower, a 19-storey office tower and an underground parking for the development.

The work scope of the 88,630 m² existing structure includes design of the tower's facade, interior floor plan layouts and interior design of lobbies and common areas. Santander is the anchor tenant of that office tower.

Another 32,000 m² existing structure (Block B) will be fitted out as office building and theater.

这个综合项目包括一个大型商场及两栋办公大厦的新建工程，对现有布局的立面及室内设计进行整修，计划将其设计成一个办公大厦。其总建筑面积为 105 701 平方米，包括 66 183 平方米的 Lguatemi 商场、一座 22 层的办公大厦、一座 19 层的办公大厦及一个地下停车场。

对占地 88 630 平方米的办公楼的设计范围包括大厦的门面设计、室内平面布置图的布局以及大厅和共用区的室内设计。Santander 是这座办公大厦的主要租客。

另外一座占地 32 000 平方米的现有楼栋（B 幢）将用做办公楼和剧院。

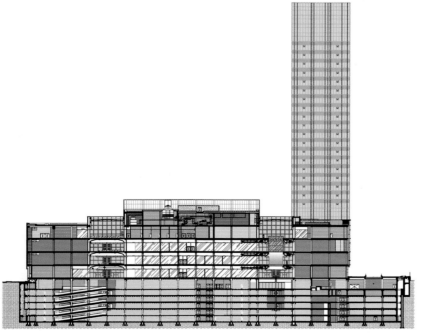

Section 1 剖面图 1

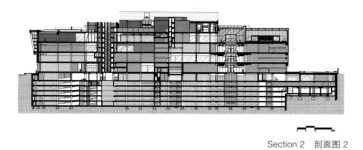

Section 2　剖面图 2

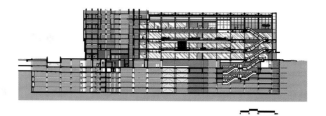

Section 3　剖面图 3

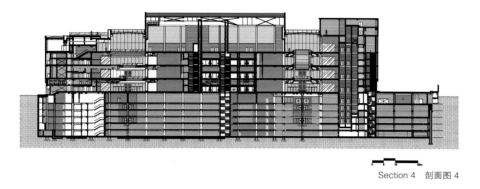

Section 4　剖面图 4

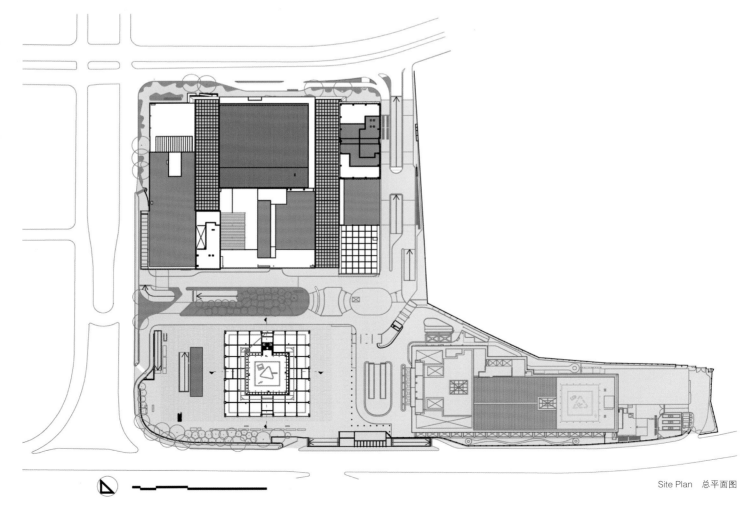

Site Plan　总平面图

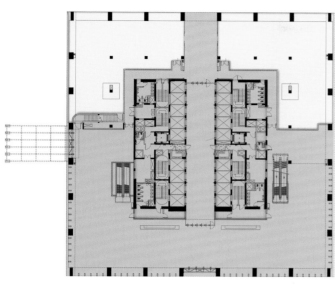

Ground Floor Plan　底层平面图

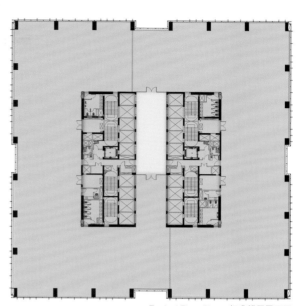

Typical Floor Plan　标准楼层平面图

ZHUHAI HUARONG HENGQIN TOWER
珠海华融横琴大厦

Architects: Atkins
Location: Zhuhai
Area: 99,631 m²

设计机构：Atkins 公司
项目地点：中国珠海市
面积：99 631 平方米

Atkins' Architecture and Urban Design Studio are pleased to announce their appointment for the design and construction of Huarong Hengqin Tower in Zhuhai. Atkins' Architecture and Urban Design Studio were selected by the developer, Huarong Real Estate Ltd. Co. through design competition, as having provided an eye-catching and modern design which specifically responds to the site, the environment, and Huarong's design aspirations.

Design Director, Mr. Ian Milne noted: "The organic towers of this project are carefully arranged to provide hotel guests and office users with stunning views of significant landmarks in Macau, while respecting the development potential and view corridors of adjacent sites. Our design is environmentally efficient and sustainable, while maximizing the development potential and inter-connectivity of the site."

This project is the first appointment of Atkins' Architecture and Urban Design Studio by Huarong, and with Huarong's solid power and strength in the Chinese market, Huarong Hengqin Tower will become the latest landmark in Zhuhai.

Huarong Hengqin Tower in Zhuhai is a new mixed-use development, it will be one of the first buildings to be built as part of the large masterplan of the new special economic zone of Hengqin Island in Zhuhai. It benefits from a prime waterfront location facing the Cotai strip in Macau.

This proposal contains accommodation for a 5-star hotel, international grade A offices and top retail facilities. With 32 floors, it is 144 m tall. With 22 floors, it is 100 m tall. The mix of all these high-end components will create a striking synergy making this development a true regional business and tourist destination and a vibrant place within the future development of the hengqin special economic zone.

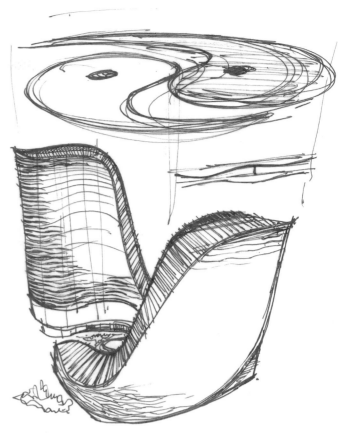

Appearance of the drawings　外观手绘图

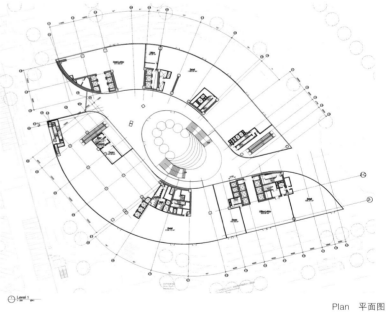

Plan　平面图

阿特金斯建筑及城市设计工作室受委托设计和建造珠海华融横琴大厦。

阿特金斯建筑及城市设计工作室依据基地现有环境条件，提供了一个醒目、现代且符合开发商华融置业有限公司愿景的设计，因此被华融选中成为中标单位。

设计总监麦意安先生表示："此项目中外形流畅生动的大厦经过精心设计，使酒店客人及办公楼用户均可以眺望到澳门地标性建筑美景。同时，大厦照顾到周边地区视野及未来的潜在发展。我们的设计不仅绿色环保，还充分提升了当地的互联性以及发展潜力。"

此项目虽然是阿特金斯建筑及城市设计与工作室与华融的首次合作，但凭借华融雄厚的企业实力，华融横琴大厦将成为珠海市最新的标志性建筑。

位于珠海横琴岛的具有综合用途的华融大厦是新经济特区首座落成的建筑物，从这里可眺望到澳门的滨海及金光大道的优美景色。建筑功能包括五星级酒店、国际甲级办公室及高级商业配套。

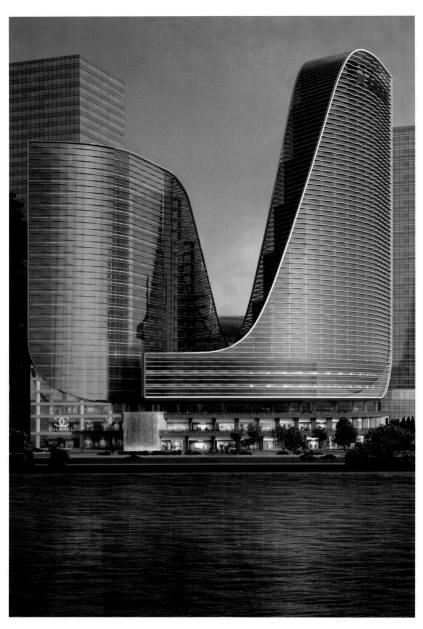

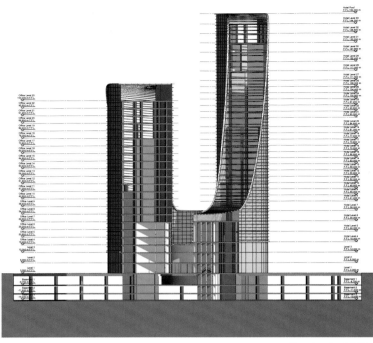

Elevation 1 立面图 1

Elevation 2 立面图 2

Modelling 1 模型图 1

Modelling 2 模型图 2

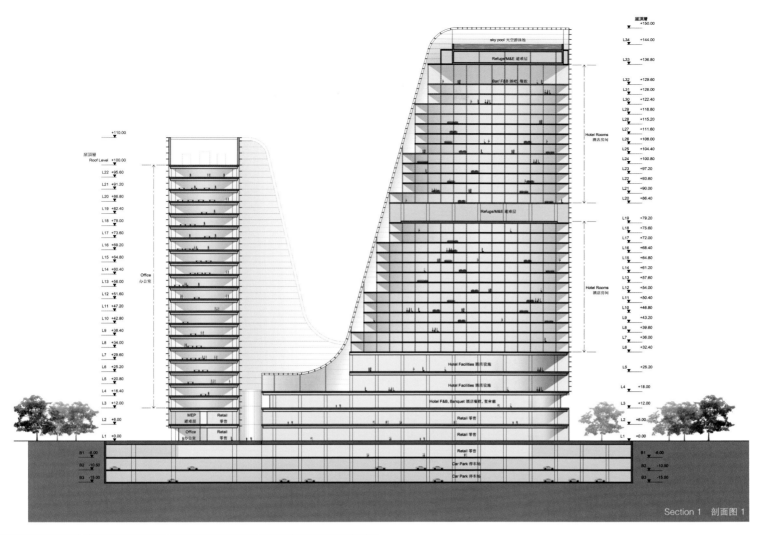

温顶窗 +150.00
L34 +144.00
L33 +136.80
L32 +129.60
L31 +126.00
L30 +122.40
L29 +118.80
L28 +115.20
L27 +111.60
L26 +108.00
L25 +104.40
L24 +100.80
L23 +97.20
L22 +93.60
L21 +90.00
L20 +86.40
L19 +79.20
L18 +75.60
L17 +72.00
L16 +68.40
L15 +64.80
L14 +61.20
L13 +57.60
L12 +54.00
L11 +50.40
L10 +46.80
L9 +43.20
L8 +39.60
L7 +36.00
L6 +32.40
L5 +25.20
L4 +18.00
L3 +12.00
L2 +6.00
L1 +0.00
B1 -6.00
B2 -10.50
B3 -15.00

+110.00
屋顶层 Roof Level +100.00
L22 +95.60
L21 +91.20
L20 +86.80
L19 +82.40
L18 +78.00
L17 +73.60
L16 +69.20
L15 +64.80
L14 +60.40
L13 +56.00
L12 +51.60
L11 +47.20
L10 +42.80
L9 +38.40
L8 +34.00
L7 +29.60
L6 +25.20
L5 +20.80
L4 +16.40
L3 +12.00
L2 +6.00
L1 +0.00
B1 -6.00
B2 -10.50
B3 -15.00

Office 办公室

sky pool 天空游泳施
Refuge/M&E 避难层
Bar/ F&B 派吧 餐饮
Hotel Rooms 酒店房间
Refuge/M&E 避难层
Hotel Rooms 酒店房间
Hotel Facilities 酒店房间施
Hotel Facilities 酒店房间施
Hotel F&B, Banquet 酒店餐饮, 宴会厅
Retail 零售
Retail 零售
Retail 零售
Car Park 停车场
Car Park 停车场

MEP 避难层 | Retail 零售
Office 办公室 | Retail 零售

Section 1 剖面图 1

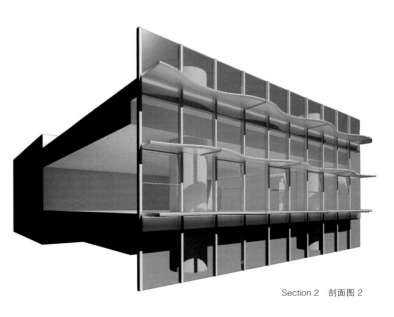

Section 2 剖面图 2

BINHAI ZHESHANG TOWER
滨海浙商大厦

Architects: ZPLUS Architecture and Planning Design Company
Chief designer: Liu Shunxiao
Location: Tianjin, China
Area: 16,800 m²

设计机构：ZPLUS 普瑞思建筑规划设计公司
首席设计师：刘顺校
项目地点：中国天津市
面积：16 800 平方米

Description of design for Tianjin Binhai Zheshang Tower
I. Project overview
This project is located in Lot 9, Xiangluowan Business Center, Tanggu District, Tianjin City, with a land area of about 16,800 m².
II. Project requirements
This project is located in the Xiangluowan Business Center and the construction standard of this project stands for the economic and cultural level of Zhejiang, so the design concept should highlight the architectural style, with novel and unique visual effect, and design for the facade effect at the main entrance is especially important. The project is expected to be a landmark in Xiangluowan Business Center after completion.
III. Introduction about design scheme
1. Concept of planning and design
1.1 Conception of two high-rise buildings
1.2 Partial subsidence at the entrance forming a rich space
The large square located in the west of the base adopts the three-dimensional design concept of partial subsidence, which enriches the landscape design and meanwhile improves the function layout, lighting and ventilation of the ground floor. Stores are arranged around the sunken plaza and people are attracted by the big step; at the same time, the underground parking lot also has better lighting and ventilation conditions. The stream of people in the parking lot can go to the first floor or other floors by directly entering the stores or vertical transportation zone. The sunken plaza is in arc shape which is vivid and the partial foot bridge increases the interest.
1.3 Traffic organization
The planned traffic organization separates people from vehicles by setting up relatively independent motor vehicle flow system and pedestrian flow system. Pedestrian flow includes different users of facilities like office, hotel, apartment and business, with each of them in their own positions. People entering the office lobby mainly start at the plaza entrance in the west of the base, pass through square fountain and green land, and then enter the public elevator hall on the 1st floor. A VIP independent hall and elevator are set for executive staffs at the northeast corner of the office building.

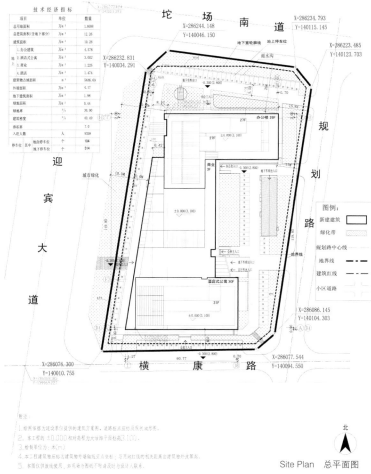

Site Plan 总平面图

People going to hotel enter the central lobby of the hotel from the south of the base, and go to the guest room floor through the cafeteria or elevator hall in the north of the lobby. People heading for the apartment enter the apartment lobby with independent hall and elevator at the southeast side of the base, while the service flow and goods flow are in a concealed place in the east. In order to increase application of commercial part, capacious stairs are set at the sunken plaza, and meanwhile escalators are established in the office and hotel lobbies, which will promote frequent use of public space above the 2nd floor. The underground garage has three entrances that are located in the north, east and west respectively; meanwhile open-parking ground is arranged at the proper place on the ground and it combines well with the green landscape.

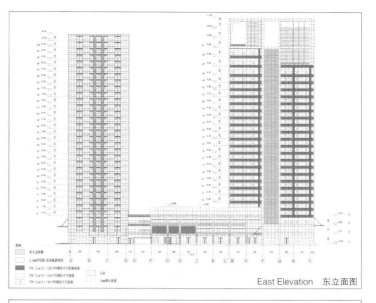

East Elevation 东立面图

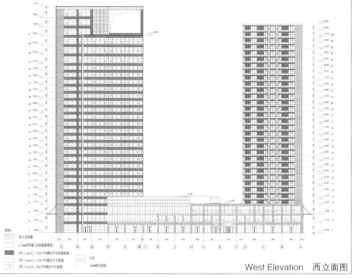

West Elevation 西立面图

天津滨海浙商大厦设计说明
一、项目概况
本项目坐落在天津市塘沽区响螺湾商务中心 9 号地块，占地面积约 16 800 平方米。
二、 项目要求
本工程地处响螺湾商务中心，该项目的设计建设标准代表着浙江省的经济、文化水平，因此要求设计立意突出建筑风格，视觉效果新颖独特，尤其考虑世纪大桥主入口的立面效果。大厦建成后有望成为响螺湾商务中心的标志性建筑之一。
三、设计方案说明
1. 规划设计概念
1.1 两座高层建筑的构思
1.2 入口广场局部下沉形成丰富的空间
位于基地西侧的大广场采用局部下沉的立体构思，丰富了景观设计，同时使地下一层的功能布局和采光通风都得以改善。在下沉广场四周布置商店，通过大台阶吸引人流，同时地下停车场也有更好的采光、通风条件。停车场的人流可直接进入商店或垂直交通区，上到一层或其他各层。
下沉广场弧形，造型活泼，局部天桥增加了趣味性。
1.3 交通组织
规划中的交通组织通过设置相对独立的机动车系流和步行系统，实现基地内的人车分流。对于步行人流，也分成办公、酒店、公寓、商业等设施的不同使用者，使之各得其所。进入办公大堂的人流主要以基地西侧的入口广场途经广场喷泉、绿地，进入首层的公共电梯厅。在办公楼东北角设有一 VIP 独立门厅和电梯，供主管人员出入。
酒店人流由基地南侧进入，通过酒店中央大堂，经大堂北部的自助餐厅或电梯厅上到客房层。
公寓人流由基地东南侧的公寓大堂进入，有独立门厅、电梯厅，服务性人流、货流处在东侧较为隐蔽之处。为提高商业部分的使用，下沉广场处有宽敞的楼梯，同时在办公楼和酒店大堂内都设有自动扶梯，促进二层以上公共部分的频繁使用。
地下车库设三个出入口，分别在基地北侧、东侧和西侧，同时在地面上的在适当位置布置了露天停车场，与绿化景观相结合。

ROCCO FORTE, ABU DHABI,UAE

阿联酋阿布扎比
Rocco Forte 综合体项目

Architects: Atkins
Location: Abu Dhabi, UAE
Client:Fa Al Farida

设计机构：Atkins 公司
项目地点：阿联酋阿布扎比
客户：Fa Al Farida

Atkins was approached to develop conceptual design and produce preliminary design, detailed design drawings and tender documentation as well as supervision services of this prestigious mixed-use development. Specifically we were involved in architecture, MEP, structural and interior design services.

Located on plot 1, Sector West 68 in Abu Dhabi within easy reach of downtown Abu Dhabi and the international airport, this mixed-use development consists of a 5-star hotel, and a luxury brand retail mall. The Rocco Forte hotel represents the first time this brand has ventured into the Middle East.

The hotel consists of a triple height main reception with ten typical floors above. A large podium sits adjacent consisting of a retail area on the ground floor, and the hotel's ballroom, business centre with meeting rooms, spa, and restaurants. A large roof terrace sits on top of the podium with green landscaped areas, an outdoor swimming pool featuring BBQ pits, a pool bar and outdoor sitting areas.

The architecture is contemporary in the form of a long fluid glass structure. Its impressive elevation is a harmonious play of coloured glass mosaic creating a dynamic flowing skin, combining various lively colours and tones. The mosaic effect along with the winding glass elevations break up the big building mass and create a dynamic elevation.

An 11-storey high glass atrium at one end of the hotel building is the main feature, housing an all-day dining restaurant. The atrium encloses an oyster shaped bar suspended at mid-height on the sixth floor.

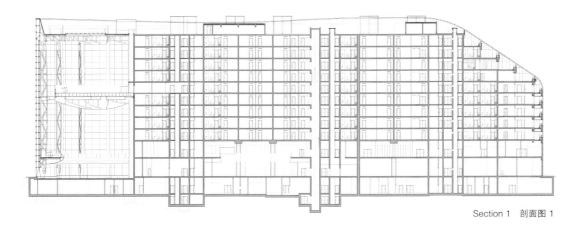

Section 1 剖面图 1

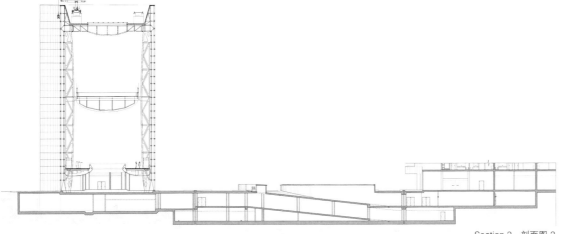

在 Rocco Forte 综合体项目中，Atkins 公司承担方案设计、扩初设计、施工图设计及工程监管等任务。确切地说，它是这个项目的总承包方，承包了建筑设计、机电设计、结构设计及室内设计。

项目位于阿布扎比西区 68 号 1 号地块，临近市区，距离国际机场仅 15 分钟的车程。Rocco Forte 综合体项目包括一个五星级酒店和一个高级品牌购物中心。作为第一座驻扎中东地区的 Rocco Forte 酒店，它的建成具有里程碑式的意义。

3 层通透挑高的大堂和 10 层舒适的酒店标准层设计展示了酒店的奢华大气。其配套裙房由高级品牌购物区、宴会厅、商务会议中心、水疗中心和餐厅组成。裙房顶部的露台除了提供景观绿化，更配备了室外泳池、BBQ 烧烤区、酒吧等一系列休闲设施，为酒店客户提供奢华的享受。

酒店的外观就像是一座巨大的玻璃建筑。这是一座非常奢华的酒店，我们的设计使用弯曲玻璃打造整座建筑。从该建筑的外观可以看出，整座建筑没有任何直线设计。玻璃的运用使得建筑外观呈现出蓝绿交织的迷人效果，也成为这片区域一道亮丽的风景线。

酒店的主要特征是在其一端配置了一个 11 层通高的餐饮中庭。牡蛎形状、曲线优美的酒吧设置在中庭空间中，看上去就好像一只牡蛎悠闲地浮在空中，给人各种不同的视觉与空间体验。

Section 2 剖面图 2

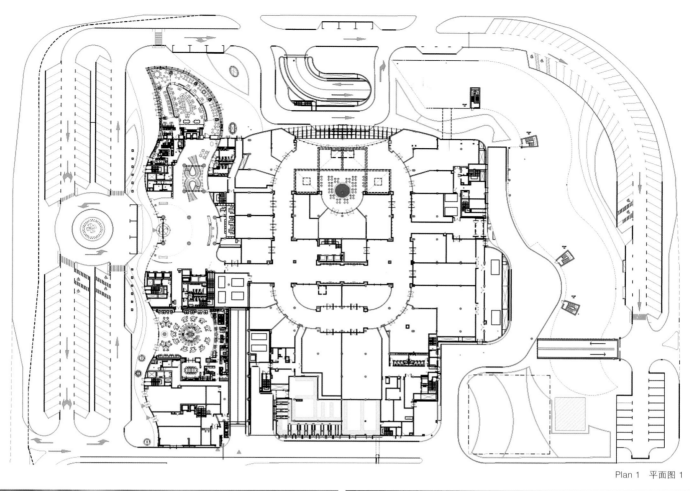

Plan 1　平面图 1

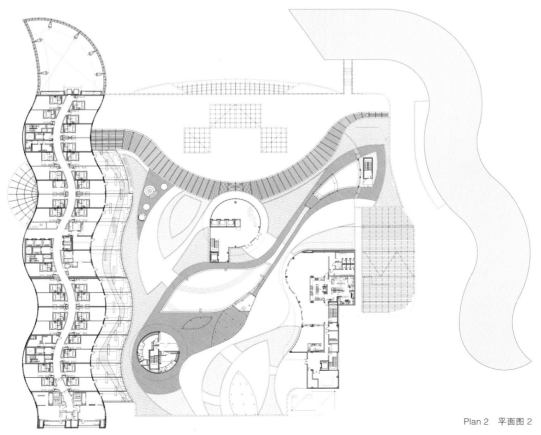

Plan 2　平面图 2

TAIYUAN WUSU INTERNATIONAL AIRPORT
太原武宿国际机场

Architecs: Atkins China
Location: Taiyuan, China
Area: 55,000 m²

设计机构：Atkins 中国公司
项目地点：中国太原市
面积：55 000 平方米

Taiyuan Wuxu Airport is located in the southeast of Taiyuan which is the capital city of Shanxi Province and its political,economic and cultural centre. Its history is long. In ancient times, Taiyuan was an important military town but it is now one of China's most important centres of heavy industry, using more than half the nation's coal output.

Fundamental to the planning for this new terminal was maintaining the efficient relationship between the terminal and the flight area. Also, the terminal zone had to satisfy the demands of both short term and long term operation. A building form angled at 45 degrees is used to develop the character of the site. By joining the new main building and its gate piers to the existing terminal, it makes it a visual as well as a functional part of the entire complex.

The existing building and the new extension merge around three courtyards that are evocative of traditional Shanxi courtyards. The courtyards are open internal spaces bringing welcoming light and air into the centre of the plan where they can be appreciated by domestic and international passengers which they separate.

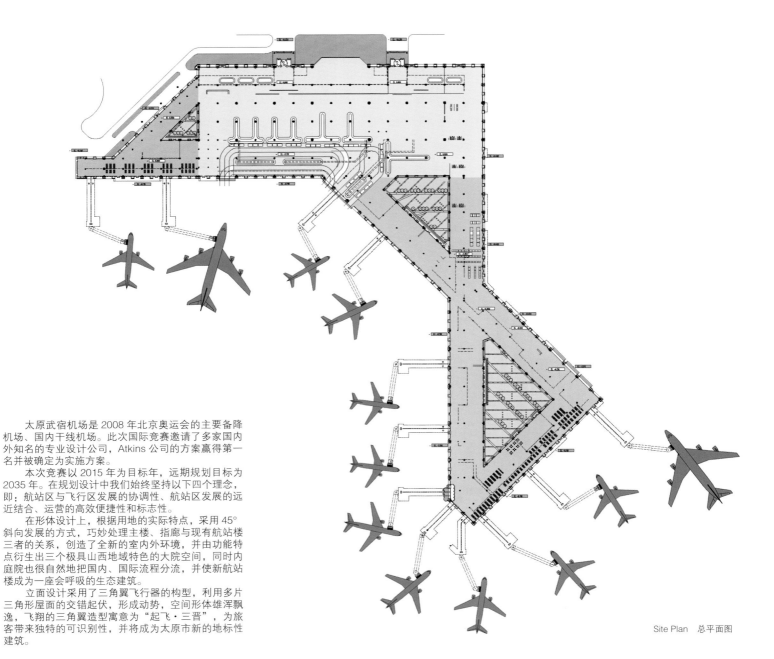

太原武宿机场是 2008 年北京奥运会的主要备降机场、国内干线机场。此次国际竞赛邀请了多家国内外知名的专业设计公司，Atkins 公司的方案赢得第一名并被确定为实施方案。

本次竞赛以 2015 年为目标年，远期规划目标为 2035 年。在规划设计中我们始终坚持以下四个理念，即：航站区与飞行区发展的协调性、航站区发展的远近结合、运营的高效便捷性和标志性。

在形体设计上，根据用地的实际特点，采用 45°斜向发展的方式，巧妙处理主楼、指廊与现有航站楼三者的关系，创造了全新的室内外环境，并由功能特点衍生出三个极具山西地域特色的大院空间，同时内庭院也很自然地把国内、国际流程分流，并使新航站楼成为一座会呼吸的生态建筑。

立面设计采用了三角翼飞行器的构型，利用多片三角形屋面的交错起伏，形成动势，空间形体雄浑飘逸，飞翔的三角翼造型寓意为"起飞·三晋"，为旅客带来独特的可识别性，并将成为太原市新的地标性建筑。

Site Plan 总平面图

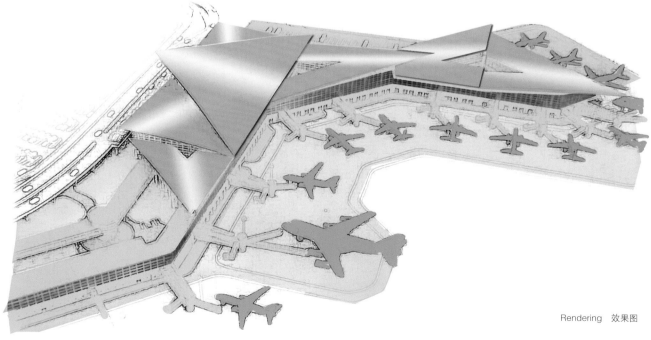

Rendering 效果图

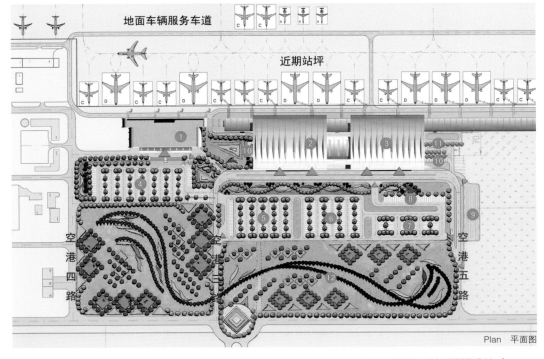

地面车辆服务车道

近期站坪

空港四路

空港五路

Plan 平面图

WANJING INTERNATIONAL BUILDING
万景国际

Architects: US DF International (Shenzhen) Architecture Design Co., Ltd.
Location: Chengdu, China
Area: 170,000 m²

设计机构：美国 DF 国际（深圳）建筑设计有限公司
项目地点：中国成都市
面积：170 000 平方米

This project is a commercial and residential complex in Chengdu. The planning and design cover two parts: the commercial part is located at the intersection and the residential block with functions like chamber is relatively close to the inner district. When this project was designed, two principal contradictions had to be solved: one was to reasonably balance the contradictory relation between residential function and commercial function under the requirement of high plot ratio and maximize the interest as well as minimize their mutual influence; the other one was to balance correlations inside the residence, and maximize the commercial value of residence by fully taking advantage of waterscape in the east and landscape in the community on the premise of satisfying requirements of sunlight. The facade is novel, fashionable and changeable; the color is rich but not messy; it highlights epochal character and symbolic character, giving people a new feeling of vigor, brightness and charm. Therefore, economic value and taste of the commercial and residential areas have been promoted on the whole and the project will be the core of new urban commerce.

本项目是成都的一个商住综合体，规划设计共分为两大部分：处于交叉口位置的地块设计为商业部分；处于相对偏里位置处的地块设计为住宅用地，住宅用地内会配套相关的会所等功能。我们设计这个项目时着力解决两个主要矛盾：一是在高容积率的要求之下，合理平衡好住宅与商业的两大功能之间的矛盾关系，使其均能达到利益最大化且使相互间的影响降到最小；二是平衡好住宅内部的相互关系，在尽量满足日照朝向要求的同时，充分利用东面的江面景观和小区内部景观，以求使住宅的商业价值达到最大。立面设计新颖、时尚、造型多变，色彩丰富而不凌乱，凸显时代性、标志性，给人以生机勃勃、特色鲜明、魅力独特的全新感觉，从而整体上提升了商业和住宅的经济价值和品位，项目建成后将成为新的城市商业核心。

SHANGHAI ZHANGJIANG CENTRAL DISTRICT URBAN DESIGN PLAN
上海张江中区城市设计

Architects: JWDA
Location : Shang Hai, China
Area : 2,100,000 m²

设计机构：JWDA 骏地设计
项目地点：中国上海市
面积：2 100 000 平方米

Running south from Zhangjiang High-tech Park, Zhangjiang Central District is expected to blossom into another high-tech corridor that brings an influx of businesses and services while ensuring the spatial diffusion of industry, commerce and residence by accommodating population growth and lifestyle needs. The district is envisioned as a campus of mixed-use functionalities featuring Chinese Academy of Sciences, institutes of higher education, financial and business facilities, low density research centers, the Knowledge Island, cultural amenities, as well as accessible public transportation. As a key node on the axis of Pudong development, it heralds a regional community center with shared services radiating out to its neighboring townships.

By combing through the fabric of spatial relationships, circulation patterns, and open spaces, the design instills a genuine sense of community, humanity and sustainability. The overall design concept embodies a commitment by Zhangjiang Central District master planning to leverage functionalities and urban planning with an eye to the character of humanism. Shared open spaces, such as streets, parks, plazas, and landscape, are strategically planned to create a vibrant Zhangjiang Central District that ignites inspiration and empowers innovation.

Urban design objectives cultivate a vital scientific community to promote interpersonal communication and cooperation, foster ecological sustainability , emphasize healthy lifestyles

Site Plan 总平面图

作为张江高科技园区的南部扩展区域,张江中区主要包括中科院、大学城、核心商业商务、低密度科研办公、智慧岛等几大功能区块。这一区域将成为园区的中心商务服务区,融合园区产业发展、人口导入、产业配套和生活配套四者的综合需求,承担起区域服务的功能。同时,作为浦东开发轴线上的一个节点,张江中区也将成为辐射周边区域的服务中心,承担起城市区域中心的功能。区域内将配备完善的科研办公、商业服务、公共文化设施与便捷的交通连接。本设计通过组织空间结构、交通结构和开放空间来促进人与人的交流与合作,提供现代的工作和生活方式,促进生态与可持续发展,创造一个充满活力与创造激情的高科技产业服务社区。

以城市设计为基础,设计师提出了城市设计导则,用于指导区域内各地块、科研办公、教育、公共设施、商业等项目的设计与建设,同时还将用于指导街道、绿化、公园、广场等城市公共开放空间的设计与建设,从而创造出充满活力与创造激情的张江中区。

SHIMAO CHENGDU
CHENGHUA COMPLEX
世茂成都成华综合体

Architecs: DAO
Location: Chengdu
Area: 2,640,000 m²

设计机构：DAO 国际设计集团
项目地点：中国成都市
面积：2 640 000 平方米

The site is distributed to focus on creating a commercial and cultural center. The overall design has ado pted the concept of "3E" — Explore, Eliminate, Entertainment to introduce a new life style, transit the city into the global financial center and provide various entertainment activities. The architect also adopted the concept of NEW — Nature, Exchange and water to provide the users a natural greenery area, and eliminate the interference between the passenger flow and the traffic flow. All functions and connections are spread out from the twin tower located in the center of the block, like a blossom hibiscus flower corresponding to the name of the city "Hibiscus City". The ferris wheel which enables the visitors to overlook the site is located on the eastern side of the twin tower. It is the landmark in this project.

成都成华总体规划关注于创造一个商业与文化娱乐中心。整体设计运用"3E — Explore, Eliminate, Entertainment，开拓新的生活方式，转变成为全球金融中心，提供万花筒般多样的娱乐活动"。建筑师还利用"NEW 概念 — Nature, Exchange, Water 为使用者提供自然绿色遮荫，并消除行人与车辆交通的互相干扰。所有的功能及联系都从中心区双子塔中心绿轴向外展开，犹如正在盛开的芙蓉花，并与"蓉城"之名相呼应。可以俯瞰地块的摩天轮位于双子塔基地东面，形成该项目的地标。

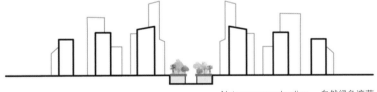

Nature green shading   自然绿色遮荫

Avoid traffic from pedestrian friendly area   利用人行区减速减缓繁忙交通

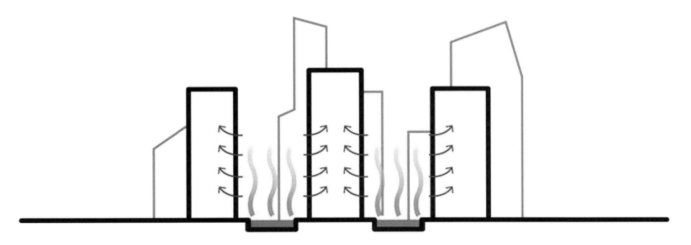

Water vaporation cooling strategy   水蒸发冷却系统

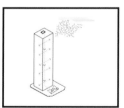

Rainwater collection
雨水收集

Sunlight reflected to ground
阳光反射至人行高度

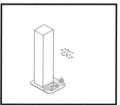

Green space to encourage habitats
地面绿化空间带来生态多样性

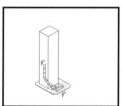

Combined heating and power
基地供电与供暖结合

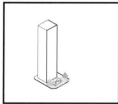

Tower allows for green space
塔楼为绿化提供空间

Sky space for social interaction
空中花园提供社交场所

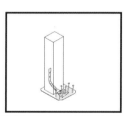

Ground scrape to act as heat sink
设备产生热源为地面供暖

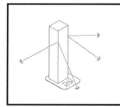

Orientation to maximum view
提供景观资源的最佳朝向

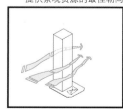

Orientated to cater to wind
塔楼根据主导风向布置朝向

Analysis　分析图

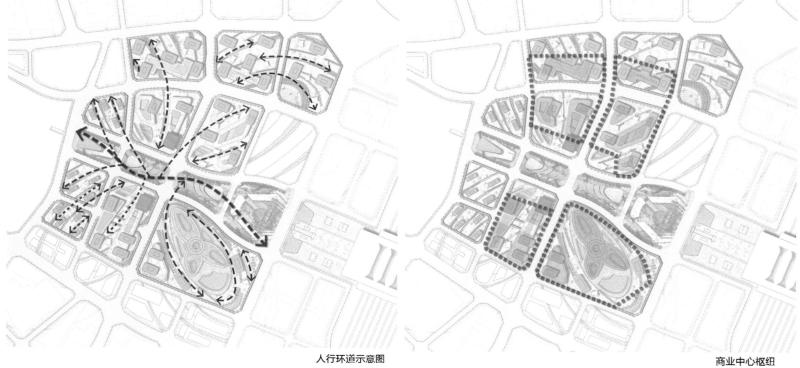

人行环道示意图
PEDESTRIAN CIRCULATION DIAGRAM

商业中心枢纽
CENTRAL RETAIL HUBS

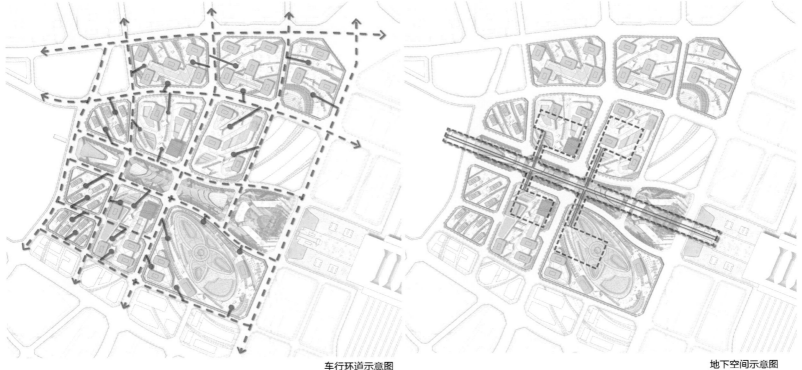

车行环道示意图
VEHICULAR CIRCULATION DIAGRAM

地下空间示意图
UNDERGROUND SPACE DIAGRAM

DOCKS EN SEINE
塞纳河码头

Architects: Jakob + Macfarlane
Lighting : Yann Kersalé
Area: 20,000 m²
Location: Paris, France
Photographer: Nicolas Borel

设计机构：Jakob + Macfarlane
灯光设计：Yann Kersal é
面积：20 000 平方米
项目地点：法国巴黎市
摄影：Nicolas Borel

The existing structure was built in 1907 as an industrial warehouse facility for the Port of Paris and was the first reinforced concrete building in Paris. The 3-storey structure was conceived as a series of 4 pavilions, each with one 10 m wide bay and four 7.5 m wide bays. On the level corresponding to Quai Austerlitz, the 10 m bay is accessible, facilitating the storing and loading of materials for transport.

The idea of the new project was to create a new external skin that is inspired primarily by the flux of Seine. The skin both protects the existing structure and forms a new layer containing most of the public circulation systems and added program, as well as creating a new top floor for the existing building. An arbor generating method is used to create a new system from the existing system, that is, "growing" a new building from the old as new branches grow on a tree. This skin is created principally from glass exterior skin, steel structure, wood deck and grassed, faceted roofscape. The programme is a rich mix centred on the themes of design and fashion, including exhibition spaces, the French Fashion Institute, music studios, bookshops, cafes, and a restaurant.

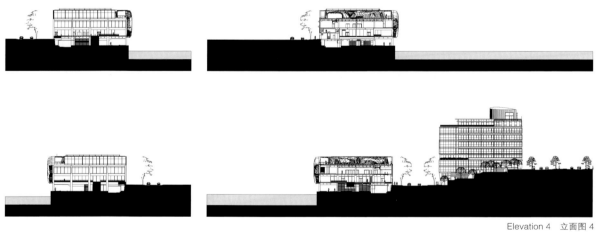

Elevation 4　立面图 4

Elevation 1　立面图 1

Elevation 2　立面图 2

Elevation 3　立面图 3

现有建筑完成于1907年，一直以来，它被用做巴黎港口的工业仓库设施，而且是巴黎首座钢筋混凝土建筑。10米宽的湖湾处于与奎奥斯特利兹河流相对的水平位置，交通便利，易于储存和装卸材料。

新项目的构想是设计一层新的表皮，最初的灵感来源于塞纳河的变迁。表皮不但起到保护现有建筑的作用，并且形成新的层次，使之可以容纳大部分的公共流通系统，同时也为现有建筑增设了顶楼。采用"植树"的方式由现有系统创设新系统，这就犹如旧建筑"长出"新建筑，和"大树长新枝"是一个概念。这层表皮大体上由玻璃外墙表皮、钢结构、木甲板和小平面的草坪式楼顶景观组成。该项目具有丰富的"混搭"风格，围绕设计和时尚的主题展开，包括展示空间、法国时装研究协会、音乐录制工作室、书店、咖啡厅以及餐厅。

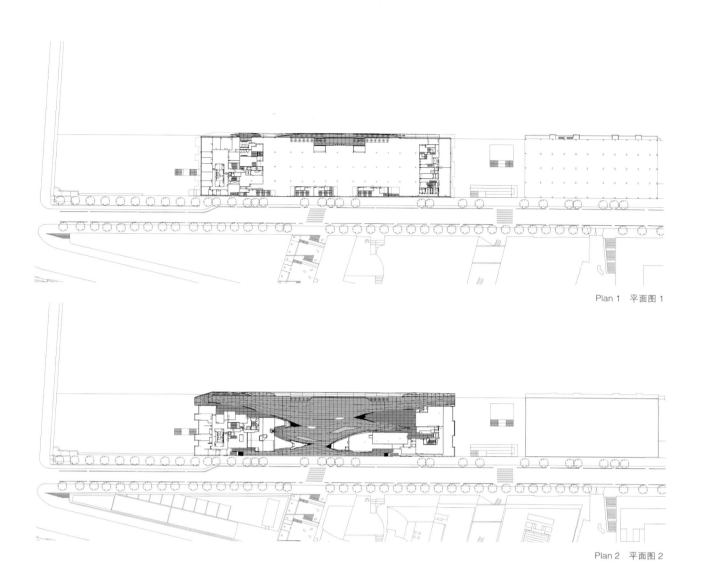

Plan 1　平面图 1

Plan 2　平面图 2

文化建筑 CULTURAL BUILDING

Being a kind of human behavior and way of life, culture is the bearer of the fruit of human civilization as well as an embodiment of the movement of civilization itself. Urban culture generally includes three parts: the material culture which represents the sensatory recognition of the city image, the technology culture which represents the spiritual identification of the city, and the norm culture which represents the regulatory distinction. It is the spirit and soul of the city. Any public cultural building can be considered as the material carrier of the spirit of the city, and the most important way to express its urban culture. However, the expression of architecture is showed by its model and space, and is melted into the architectural environment through the perception of historic culture and traditions.

When designers design public cultural buildings, they should integrate a variety of complicated factors together into a coherent whole and make a appropriate choice on the premise of comprehensive judgment. After the beforehand studies and summaries of design work, designers can begin with three aspects. Firstly, start with theme.This case usually aims at cultural displays and memorial sites for a specific topic and is always provided with relatively specific cultural performances and clear psychological anticipation on scale. Secondly, start with site environment. Many cultural buildings' cultural connotations are not provided beforehand, but depend on particular circumstances. In this case designers should proceed from the specific nature and human environments of the projects to extract useful information and manifest the true cultural connotations of buildings. Thirdly, start with contemporary values and aesthetics. With the continual appearance of new structures, new materials, new technologies, contemporary architectural forms are provided with protean manifestations. The emphasis of eco-energy saving, sustainability, merging and coexistence of public buildings and urban life, as well as sophisticated and variable building skins and rich design methods of contemporary public cultural buildings are the most distinctive brand of the age for contemporary public cultural buildings.

Culture not only provides architecture with vitality, but also is spread widely by the expression of architecture. When building is provided with a cultural connotation, it is no longer a reinforced concrete pouring, but a cultural symbol of a city or a country.

　　文化是人类的一种行为方式和生存方式，是人类文明成果的承载，也是文明本体运动的体现。城市文化一般是由代表城市形象感观识别的物质文化、代表城市精神识别的技术文化以及代表规范识别的规范文化等组成，它是一个城市的精神与灵魂。任何一座公共文化建筑，都是城市精神的物质载体，是城市文化最重要的表达方式。而建筑的这种表达是通过造型和空间表现出来的，并通过对历史文化和传统的感悟形成认识，将之融入建筑环境之中。

　　文化是人类的一种行为方式和生存方式，是人类文明成果的承载，也是文明本体运动的体现。城市文化一般是由代表城市形象感观识别的物质文化、代表城市精神识别的技术文化以及代表规范识别的规范文化等组成，是一个城市的精神与灵魂。任何一座公共文化建筑，都是城市精神的物质载体，是城市文化最重要的表达方式。而建筑的这种表达是通过造型和空间表现出来的，并通过对历史文化和传统的感悟形成认识，将之融入建筑环境之中。

　　设计师在设计公共文化建筑的时候，需要统筹考虑各种复杂因素，并在综合判断的基础上做出适宜的选择。经过对前期工作的设计研究和汇总，设计师可以从三个方面来切入：首先，以主题世界为切入点，这主要是针对某一具体的主题实践的文化展示和纪念场所而言，在这种情况下，公共建筑通常会拥有比较明确的文化性表现，对尺度也有明确的心理预期；其次，以场地环境为切入点，很多公共文化建筑所要呈现的文化内涵都不是事先给定的，这需要设计者从项目自身的具体情况出发，根据场地的自然环境和人文环境状况，提炼出有用的信息，表达出建筑所应有的文化内涵；第三，以当代价值观及审美观为切入点，伴随着新结构、新材料、新技术的不断涌现，当代建筑形式有了千变万化的表现，而强调建筑的生态节能和可持续发展，公共建筑与城市生活的共生融合以及复杂多变的建筑表皮，丰富多彩的创作设计，都为当代公共文化建筑烙下了最鲜明的时代烙印。

　　文化不仅赋予建筑以生命力，同时也通过建筑的表达而得到了加深与传播。建筑有了文化内涵，就不再是单纯的钢筋混凝土浇筑物，而是一个城市乃至一个国家的文化性标志。

THE OCEAN SURF MUSEUM
海洋冲浪博物馆

Architects: Steven Holl Architects
Location: Biarritz, France
Area: 3,800 m²
Photographer: Iwan Baan

设计机构：Steven Holl 建筑事务所
项目地点：法国比亚里茨市
面积：3 800 平方米
摄影：Iwan Baan

The ocean surf museum is a museum that explores both surf and sea and their role upon our leisure, science and ecology. The design by Steven Holl Architects in collaboration with Solange Fabia is winning scheme from an international competition in 2005 that included the offices of Enric Miralles/Benedetta Tagliabue, Brochet Lajus Pueyo, Bernard Tschumi and Jean-Michel Willmotte. The building form derives from the spatial concept "under the sky" , "under the sea" .A concave "under the sky" shape forms the character of the main exterior plaza, the"Place de l'Ocen", which is open to the sky and sea, with the horizon in the distance. A convex structural ceiling forms the "under the sea" exhibition spaces. Two "glass boulders", which contain the restaurant and the surfers' kiosk, activate the central outdoor plaza and connect analogically to the two great boulders on the beach in the distance. The building's southwest corner is dedicated to the surfers' hangout with a skate pool at the top and an open porch underneath that connects to the auditorium and exhibition spaces inside the museum. This covered area provides a sheltered space for outdoor interaction, meetings and events.The gardens of The ocean surf museum aim at a fusion of architecture and landscape, connecting the project to the ocean horizon. The precise integration of concept and topography gives the building its unique profile. The materials of the public plaza are progressively vary Portuguese cobblestone with grass and natural vegetation. Towards the ocean, the concave form of the building plaza is extended through the landscape. With slightly cupped edges, these gardens,a mix of field and local vegetation, are a continuation of the building and will host the idea of the architecture design.

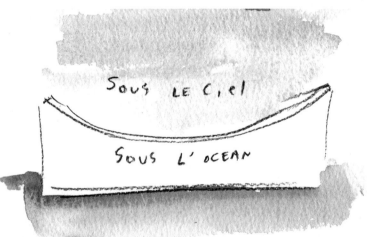

SOUS LE Ciel

SOUS L' OCEAN

PLACE de L'OCEAN, BIARRitz

海洋冲浪博物馆是一个展示冲浪和海洋以及它们在我们的休闲、科学和生态学中的角色的博物馆。建筑设计由 Steven Holl 建筑工作室与 Solange Fabia 合作完成。他们在 2005 年的国际竞标活动中脱颖而出，一同竞标的还有 Enric Miralles 和 Benedetta Tagliabue 建筑师事务所、Brochet Lajus Pueyo 建筑师事务所以及 Bernard Tschumi 和 Jean-Michel Willmotte 建筑师事务所。建筑的形式来源于"在天空下"和"在海洋下"的概念设计。"在天空下"部分的一个凹面形成了外部广场和海浪冲浪博物馆的主要特点，在远处的水平面上对天空和大海敞开。一个凸形的天花板形成了"在海洋下"的展览空间。两个包含餐厅和冲浪者凉亭的"玻璃巨石"活跃了中央室外广场，并且和远处海滩上两块巨大的岩石相联系。建筑的西南角是供冲浪者玩耍的顶部滑冰场和连接礼堂与博物馆内的展览空间的开放门廊。这个覆顶的区域提供了一个户外交流、会议和议事的庇护空间。海洋冲浪博物馆的花园旨在使建筑和空间融合，将建筑场地和海平面连接起来。概念和地势的巧妙结合给予建筑一个独特的外形。公共广场的材料是由草和天然植被覆盖的渐进变化的葡萄牙鹅卵石。面对海洋，建筑广场的凹面形式通过景观而延伸。具有微凹边缘的花园、旷野和植被的组合是建筑设计理念的延续。

Profile drawings　剖面手绘图

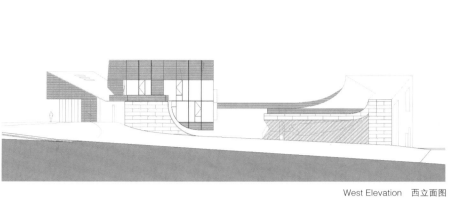

West Elevation　西立面图

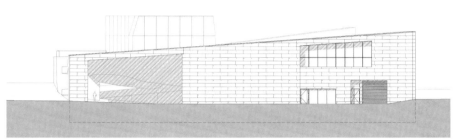

East Elevation　东立面图

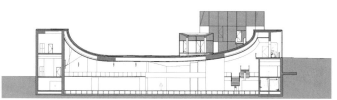

Section 1　剖面图 1

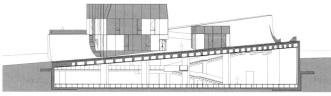

Section 2　剖面图 2

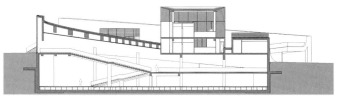

Section 3　剖面图 3

Section 4　剖面图 4

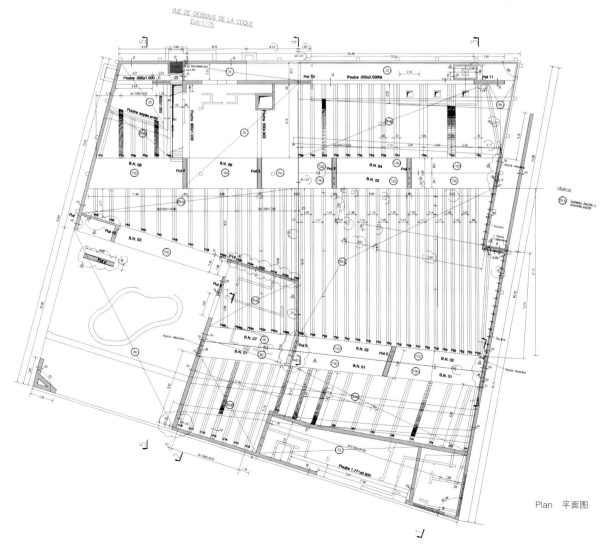

VUE DE DESSOUS DE LA COQUE
Echt:1/75

Plan　平面图

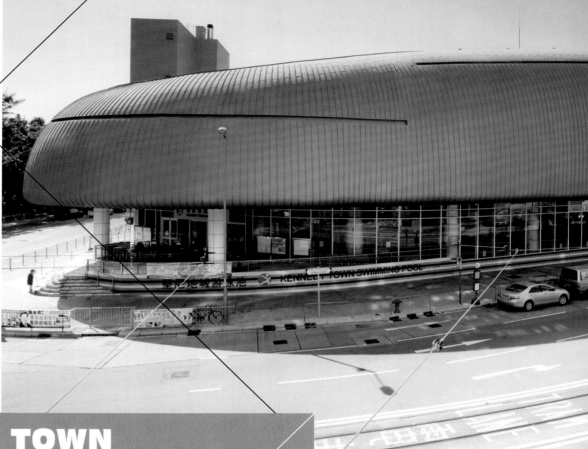

KENNEDY TOWN SWIMMING POOL

坚尼地城游泳池

Architects: TFP Farrells
Client: Mass Transit Railway Corporation Limited
Location: Hong Kong
Site Area: 8,842 m²

设计机构：TFP Farrells
项目地点：香港
面积：8 842 平方米
客户：香港地铁有限公司

Re-provisioning of Kennedy Town Swimming Pool is one of several enabling work project on MTRCL's West Island Line development. The new swimming pool complex is planned to be constructed in two phases to facilitate concurrent development and early opening of the new outdoor pools. One of the conditions imposed to the MTRCL's resumption of the existing pool site is that the current facility should not be demolished until Phase 1 is in operation. Since May 2011, Phase 1 has been opened to the public; the remaining half of the site is now used as a temporary works area for MTRCL WIL construction.

The site of the new swimming pool complex is defined by Kennedy Town Praya, Sai Cheung Street North and Shing Sai Road. The triangular shaped complex now accommodates outdoor secondary pool, leisure pool, jacuzzi, staff rooms and filtration plant in Phase 1. Indoor multi-purpose pool, teaching pool, jacuzzi and an outdoor sitting area will be provided in Phase 2.

Situated on the reclaimed land, the new swimming pool, together with the existing residential towers and the historic tramway along the old waterfront, forms the backdrop for the new development. The low-lying, organic and singular expression of its architectural design marks the entrance to Kennedy Town as well as its prominent location next to Victoria Harbour. The ground level is designed to set back, providing a shaded queuing area and heightening the impression of the building hovering above ground.

The entrance to the complex lies at the corner of Kennedy Town Praya and Sai Cheung Street North, where the full height glazing is extended to enclose the full length of the Crush Hall, ensuring a visual connection to the adjacent historic tramway. The escalators and lifts from the Crush Hall lead to the pools, jacuzzi and changing rooms at Level 1, where swimmers and visitors may rest on the pool deck and enjoy the uninterrupted views of Victoria Harbour and Belcher Bay Park. An operable glass wall is built to separate the indoor and outdoor pools, allowing the two to be combined during the summer season.

The primary cladding material used for the building envelope is the pre-weathered zinc. The cladding forms a continuous skin that not only wraps around the outdoor pool, but also keeps going upwards to form the roof and touch down to the ground over the indoor pools, to provide a canopy over the outdoor sitting area. The natural, self healing pre-weathered zinc provides a durable and lasting finish to the building. Its low sheen minimizes reflection to neighborhood; meanwhile the strength and malleability enable a curvilinear form to be achieved.

Transparent ETFE cushions are proposed to build the roof over Phase 2 indoor swimming pools to maximize daylight and simulate an outdoor environment in the pool hall. The lightweight air filled cushions are able to achieve large spans without intermediate supports, hence minimize visual interruptions. In addition, the air filling also acts as an insulator which reduces the heat gain into the pool hall.

Phase 2 is tentatively scheduled for completion in 2016 after MTRCL's WIL comes into operation.

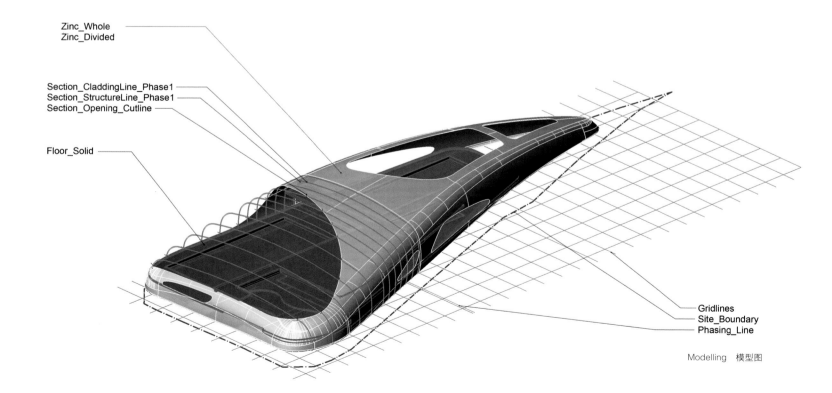

Modelling　模型图

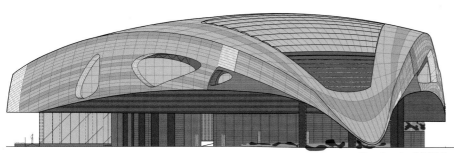

坚尼地城游泳池（Kennedy Town Swimming Pool）是香港地铁有限公司（MTRCL）西港岛线前期发展项目之一。游泳池的主体将分成两个阶段兴建，务求配合西港岛线工程同步发展，并加速多个全新室外游泳池尽早对公众开放。香港地铁有限公司（MTRCL）该项游泳池重建工程附带条件之一，是确保现有设施于全新游泳池第一阶段工程投入运作前不被拆除。2011 年 5 月，坚尼地城游泳池第一阶段已大功告成，并正式对外开放，余下第二期工程的施工场地现暂进行兴建港铁西港岛线项目。

该全新游泳池位于坚尼地城海傍、西祥街北和城西道之间的三角形区域，第一阶段工程为游泳池增设了室外副池、嬉水池、按摩池、员工休息间和机电设备。至于第二阶段工程则会兴建室内多功能泳池、教学泳池和按摩池，以及户外绿化休憩区。

全新坚尼地城游泳池坐落于填海土地之上，与沿昔日海滨长廊而建的住宅大厦及历史悠久的电车轨道一同为区内全新发展提供理想环境。游泳池地理位置优越，紧邻维多利亚港，其别出心裁的单体建筑设计流丽纤薄而靠近地面，点缀了坚尼地城的入口处，而地下层更刻意向内缩小，腾空搭建出有盖排队区域，进一步凸显其恰如巨鸟盘旋上空的感觉。

游泳池的入口处位于坚尼地城海傍与西祥街北交界，落地玻璃从大门一直延伸，包围整个收费大堂，让旁边历史悠久的电车轨道美景尽映眼帘。从收费大堂乘搭电梯或自动扶手梯即可直达一楼游泳池、按摩池和更衣室。游客及泳客站在池边，可以无限眺望维多利亚港和卑路乍湾公园风光。室外及室内游泳池仅以活动玻璃幕墙分隔。仲夏时分，只需拉开幕墙，游客及泳客就能同时兼享两者。

游泳池表层结构采用预钝化锌（Pre-weathered Zinc）为主要覆盖用料，遂令建筑覆层展现连绵不绝的特色，不但包围整个室外游泳池，更覆盖室内游泳池的屋顶和地面，为户外绿化休憩区铺上华盖。预钝化锌本身具备天然自愈能力，足令建筑表层持久耐用，其微弱光泽对周围建筑物造成甚少反射，而其坚固、柔韧的特质更成功为建筑打造平滑的曲线形态。

第二期工程选用透明 ETFE 气枕盖建室内游泳池屋顶，大幅提升室内空间的采光度，俨如户外开扬环境。轻巧的气枕能够用做盖建广阔建筑而无需任何支撑，让室内空间的视线阻碍减至最低，同时充满空气的结构还具备隔热功效，减缓游泳池室内温度的上升。

第二期工程预计将于 2016 年港铁西港岛线开通后完工启用。

Elevation 1　立面图 1

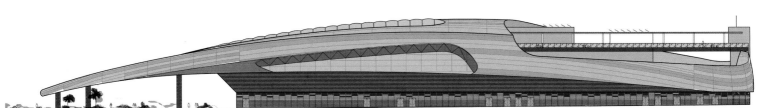

Elevation 2　立面图 2

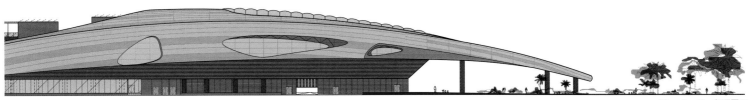

Elevation 3　立面图 3

EXPO ZARAGOZA 2008
萨拉戈萨世博会 2008

Architects: ACXT Architects
Location: Zaragoza, Spain
Area: 250,000 m²
Photographer: Aitor Ortiz

设计机构：ACXT 建筑师事务所
项目地点：西班牙萨拉戈萨市
面积：250 000 平方米
摄影：Aitor Ortiz

This project for the main area of Expo 2008, its town planning and the design of the building housing most of the exhibition area, i.e. the international pavilions and those occupied by Spanish regions, was a considerable challenge in several aspects. Firstly, because Expo 2008 is international, the specific BIE format had to be applied. This meant using the same construction concept when designing all the exhibition pavilions, and required the project to be seen as a single unit. This was an opportunity to provide Zaragoza with a first-rate building complex able to blend in with its natural and urban settings secondly, it was also a chance to design the exhibition site so that once the Expo was over it could be transformed with as little rebuilding as possible into a service and leisure area that could then be completed and consolidated as an interesting area of the city. Thirdly, the large roof not only gave the entire project a seamless appearance and image but also created an outstanding architectonic and urban identity. Finally, underpinning the entire project by the "Water and Sustainable Development" concept and theme was a driving force not only in the realm of ideas, but also as regards practical, countable questions such as energy.

The criteria applied to the arrangement of the new-build blocks clearly differentiated between those facing north near the Rabal ring road and those facing south near the river Ebro. The northern blocks provided a barrier shielding the site from the noise of traffic and the strong north wind, whilst the public zones looked straight towards the river Ebro with breath-taking views of the Basílica of Nuestra Señora del Pilar.

Another of the project's most important formal aspects was the use of organic forms inspired throughout by the fluidity of water associated with the same concept of searching for continuous, flowing spaces of a more amiable and interesting type in outdoor spaces rather reminiscent of the natural way water behaves in nature.

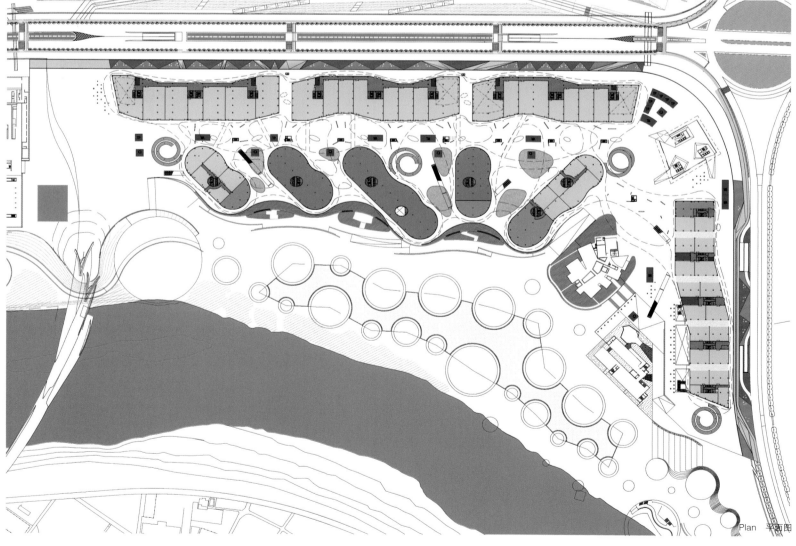

Plan 平面图

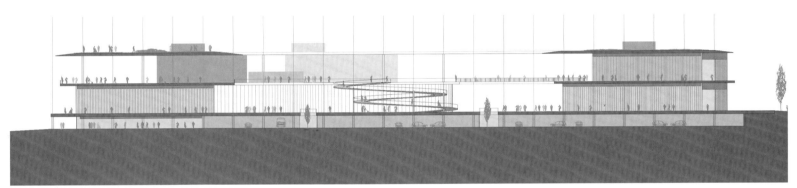

Section 1 剖面图 1

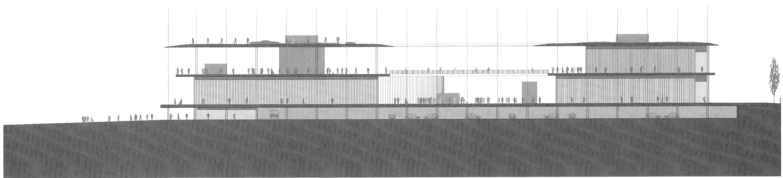

Section 2 剖面图 2

这个项目位于 2008 年世博会的主要地区，它的展区规划和建筑设计覆盖地区的大部分展区，即国际展馆和西班牙展区，面临来自于几个方面的重大挑战。首先，2008 年世博会是国际化的，因此必须采用国际展览局规定的形式，这要求在设计所有的展览场馆时，需要有相同的结构概念设计，并且需要各个建筑浑然一体，这为在萨拉戈萨建设一个一流的融合自然和城市的建筑体量提供了绝佳的机会。其次，设计展览场地也是一个挑战，因为一旦世博会结束，场地将会通过尽可能少的重建工作转变为一个服务和休闲的区域，作为一个有趣的城市区域被统一利用。再次，大型屋顶不仅仅给予整个项目一个无缝的外观和形象，而且创造了一个杰出的建筑和城市标志。最后，整个项目的基础是"水和可持续发展"这一概念主题，它不仅仅是理念领域里的推动力，而且是具有实际意义和可以量化的问题，比如能源。

新建筑的功能区清楚地区分开了面对 Rabal 环形公路的北面板块和面对 Ebro 河的南面板块。北面板块提供了一个屏障，将项目与嘈杂的交通噪声和强劲的北风隔离开来，同时公共区域直接面对 Ebro 河，可以看到震撼的 Nuestra Senora del Pilar 大教堂。

项目另一个最重要的意义是使用水的主题，这与寻求连续、流动的空间的理念相一致，寻找一种亲切的和有趣的户外空间形式，让人们回忆起水体在自然中最原始的表达方式。

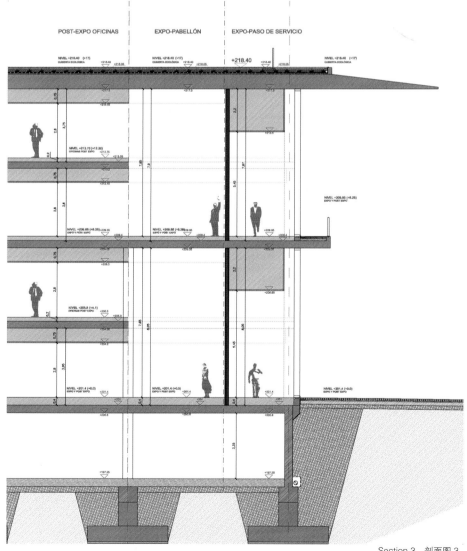

Section 3 剖面图 3

RICHMOND SCHOOL, UK
英国里士满学校

Architects: Atkins
Location: Richmond, UK
Client: North Yorkshire County Council

设计机构：Atkins 公司
项目地点：英国里士满市
客户：北约克郡议会

Richmond School campus lays approximately 2 km from the centre of Richmond, North York Shire, situated on a large site with beautiful views across the Vale of York.

The site was recognised as a school for future by a government led scheme and granted funding in order to reform and unite the school. Through this BSF funding, it is envisaged that the two disparate sites would combine and create a unified, cohesive school containing pupils from year 7 to year 13.

Atkins was commissioned to undertake an extensive master planning and re-modelling exercise of the existing Richmond school campus, which contains a diverse range of buildings, reflecting architecture trends of the last 70 years. Significant elements of the existing buildings on site have been retained, including a 1940's grade II listed building, to be used as the new 6th Form Centre for 300 students. As this building sits independently within the school campus, this helps create a separate identity for the 6th Form and enhances its status in specialist Creative Arts.

The project also includes substantial newly-built accommodation including fully equipped art rooms, science labs, music rooms, kitchen and dining facilities.

A new 2,140 m² 6-court sports hall with separate fitness suite and activity hall completes the development and provides state of the art facilities for pupils and, through Richmond Shire Leisure Trust, community use alike. The sports hall opens out onto extensive sports pitches which have been re-levelled and landscaped.

Our approach to sustainable construction encompasses every element of sustainability — including transport, food, healthy lifestyles and energy use. As a result, we successfully applied for £2 m in additional funding from the BSF programme.

Sustainable measures include:
• natural light and ventilation
• solar reflective glazing
• a sedum roof
• a bio-mass boiler fuelled with wood-chip
• grey water harvesting
• increased insulation standards for the envelope and glazing which exceed current regulations
• exposed concrete floor slabs for thermal mass benefits
• landscaped wet-land area for educational use

The "random" appearance of the window pattern is the perfect solution to avoid distracting from the adjacent listed 1940's rectilinear block, and is key to securing planning approval for this development. Close coordination was required by the design team with English Heritage to ensure design decisions meet with the rigorous requirements of building listing. The new design opens up possibilities for the school to deliver a more flexible curriculum, offer more facilities to the wider community, provide high quality working environments for staff and create state of the art learning spaces for pupils.

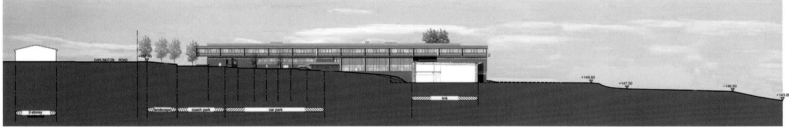

section through rectilinear building

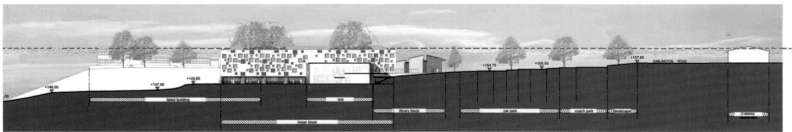

section demonstrating linear block

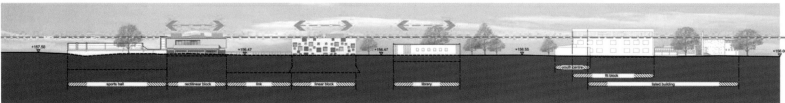

section along Darlington Road

Section 剖面图

里士满学校校园距离北约克郡里士满的市中心大约 2 千米,坐落于约克山谷中风景秀美的地区。

本项目被认定为政府规划的未来学校,这保证了对学校进行改革和资源整合的资金。利用 BSF 项目的资金,两个不同的项目将会被结合,以创造一个统一的、具有凝聚力的学校,为 7 岁到 13 岁的学生服务。

Atkins 公司被委托承担项目的总体规划和已有的里士满学校校园的重组工作,包括以不同的建筑类型来反映近 70 年来的建筑发展趋势。现有建筑的重要元素被保留下来,包括 20 世纪 40 年代的二级文物保护建筑,它将作为能容纳 300 名学生的新的六年级学生中心。因为本建筑独立于校园中,这有助于为六年级学生创造特有的分区,并且提高其在专业艺术创造中的地位。

项目还包括大量新建的学生公寓、设备齐全的艺术房间、科学实验室、音乐教室、厨房和餐饮设施。

一座占地 2 140 平方米、拥有 6 个球场和独立健身器材以及活动大厅的体育馆由里士满郡 Leisure Trust 开发,为学生提供最先进的运动设备,就像一个社区一样。位于广阔的场地上的开放运动场地已经被重新整平和美化。

我们创造可持续性建筑的方式围绕着可持续性的每一个元素,包括交通、食品、健康的生活方式和能源的使用。因此,我们从 BSF 项目计划中成功地申请了 200 万的额外资金。

可持续措施包括:
• 自然光线和通风
• 太阳能反射玻璃
• 一个种植有景天属植物的屋顶
• 一个燃烧木屑的生物锅炉
• 水体的再循环利用
• 超过现有标准的绝缘性更好的外墙和玻璃
• 具有显著热效益的混凝土楼板
• 用于教育功能的美化湿地

窗户图案的"随机"外观是避免与已有的 20 世纪 40 年代的直线区块发生冲突的完美解决方案,是确保这个发展项目的规划许可的关键。具有英国传统的设计团队需要密切的协调以得到符合建筑严格需求的设计决策。

新的设计为学校创造了开展更灵活的课程、提供更多的社区设施、为员工提供高品质的工作环境和为学生提供优良的学习空间的可能性。

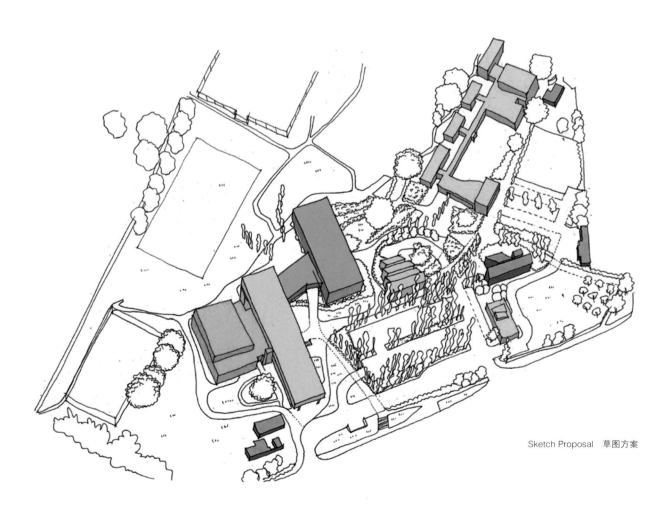

Sketch Proposal 草图方案

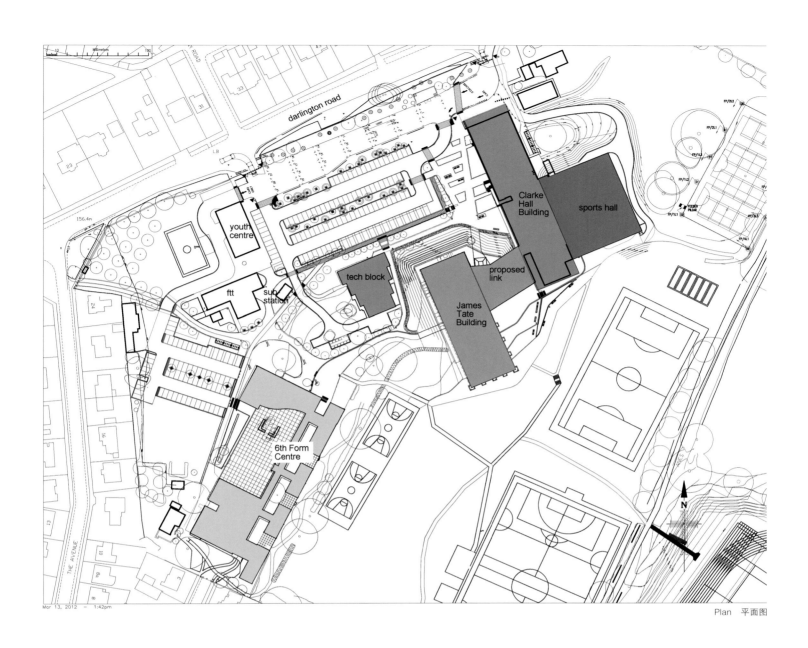

youth centre

Clarke Hall Building

sports hall

tech block

proposed link

ftt

sub station

James Tate Building

6th Form Centre

darlington road

N

Mar 13, 2012 — 1:42pm

Plan 平面图

FLORIDA INTERNATIONAL UNIVERSITY (FIU) SCHOOL OF INTERNATIONAL AND PUBLIC AFFAIRS (SIPA)

佛罗里达国际大学国际和公共事务学院

Architects: Arquitectonica
Partners-in-charge of Design: Bernardo Fort-Brescia, FAIA and Laurinda Spear, FAIA, ASLA, LEED AP
Location: Miami, USA
Area: 5,303 m²
Client: Florida International University Facilities Planning & Construction University Park
Photographer: Robin Hill, Cheryl Stieffel (courtesy New York Focus LLC.), courtesy Florida International University

设计机构：Arquitectonica
协助设计：Bernardo Fort-Brescia, FAIA 和 Laurinda Spear, FAIA, ASLA, LEED AP
项目地点：美国迈阿密市
面积：5 303 平方米
客户：佛罗里达国际大学设施规划和建设大学公园
摄影：Robin Hill, Cheryl Stieffel (courtesy New York Focus LLC.), courtesy Florida International University

A 56,000 SF (5,200 m²) building for School of International and Public Affairs is to be constructed in the main park of Florida International University. Aiming to achieve FIU's vision, the new building will provide a state-of-the-art venue for the many activities — classes, lectures, workshops, performances, conferences, and faculty and graduate student research.

The proposed structure will stand to represent the founding idea of the school as an international university, recognizing the multicultural community of Miami as the crossroads of trade, finance and culture.

Faculties from various departments will merge to advance the study of social and political sciences, international relations and humanities to foster interdisciplinary, thematic, and professional degrees and programs alongside the traditional disciplinary offerings. The building will provide a striking physical symbol of the international dimension of the university's mission and identity.

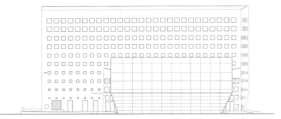

East Elevation　东立面图

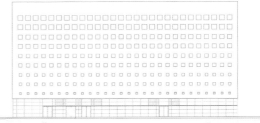

West Elevation　西立面图

South Elevation　南立面图

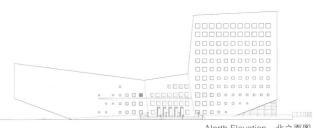

North Elevation　北立面图

　　一个5 200平方米的国际和公共事务学院建筑将要在佛罗里达国际大学的公园中建设。为了实现佛罗里达国际大学的愿景，新建筑将为很多活动，比如讲课、讲座、工作坊、演出、会议、教师和研究生的研究等提供最先进的活动场地。

　　结构设计将体现学校作为国际大学的理念，以体现迈阿密作为贸易、经济和文化交融的多元社区的地位特点。各个院系的教员将会相互融合，以推进社会和政治科学、国际关系和人文学科，促进跨学科的、主题式的专业培养，当然也包括传统培养方案的制定。建筑将成为使大学国际化的引人注目的符号和标志。

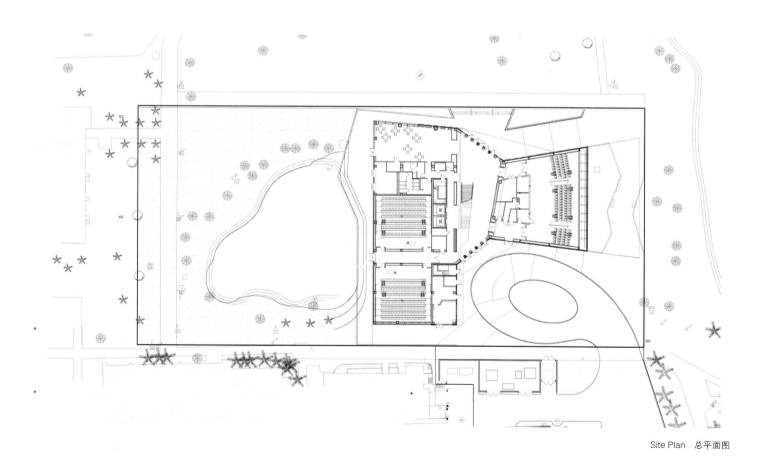

Site Plan　总平面图

ADMINISTRATION BUILDING OF CHINA UNIVERSITY OF MINING AND TECHNOLOGY

中国矿业大学行政楼

Architects: Beijing Sino-Sun Architectural Design Co., Ltd.
Location: Xuzhou, China
Area: 16,180 m²

设计机构：北京华太建筑设计有限公司
项目地点：中国徐州市
面积：16 180 平方米

The administration building of Nanhu Campus of China University of Mining and Technology is located in the east of Nanhu campus and has six floors (some part has three floors). In accordance with "practical, efficient, flexible, comfortable, beautiful, safe and economic" principle, the design tries to realize reasonable functions, flexible layout, advanced facilities, comfortable environment and convenient management.

The overall layout, shape, height, floor and color meet the requirements of campus planning, to ensure integrity of the overall style and quality of the campus. It fully reflects the principles of sustainable development and resource sharing, and introduction of ecological and humanistic concept highlights construction of a first-class university facing the new century, reflecting the economic status and characteristics of regional culture.

The administration building is located at the east entrance of Nanhu campus, which is convenient for external exchange. The buildings are arranged at the north and south side of the portal axis and squares are designed in front of the north and south entrance to form buffer spaces. This building is located in the north of the main entrance axis of the school and provides office space for leaders and faculty members in campus. On the premise of referring to the occupancy arrangement suggestions provided by the school, all office areas are re-arranged. Office space for school leaders and academicians are relatively concentrated, with separate entrances to ensure quiet and elegant working environment; office space for other faculty members is independent, with entrances facing the square before the school, to meet the working requirements. Office hall is established separately on the first floor of finance department, which is convenient for outward connection.

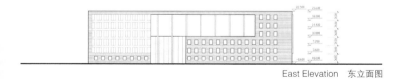

East Elevation　东立面图

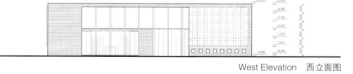

West Elevation　西立面图

South Elevation　南立面图

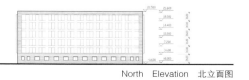

North　Elevation　北立面图

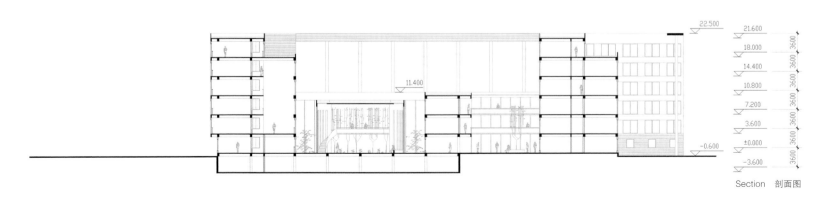

Section 剖面图

This design attaches importance on space sharing and adopts the method of lighting hall and patio space to connect office staffs at different levels together, which makes the space penetrate to each other and unite the team. Most rooms in the building adopt the layout in south-north direction, which decreases the exposure to the sun for the west side of the building by giving full consideration to the local climate that has sufficient sunlight in summer. Indoor and outdoor greening garden is utilized to form ecological climate of the building and creates a pleasant office environment. Colonnade is used to generate magnificent and elegant cultural atmosphere, maintain integrity of architectural style, reduce unnecessary ornaments, and thus make the whole building simple and generous.

中国矿业大学南湖校区校行政楼位于中国矿业大学南湖校区东侧，共有六层（局部三层）。按照"实用、高效、灵活、舒适、美观、安全、经济"的原则，设计力求功能合理、布局灵活、设施先进、环境舒适、管理便捷。

整体布局、造型、高度、层数、色彩符合校园规划的要求，确保校园整体风格的统一与品质，既充分体现可持续发展原则、资源共享原则，又引入生态和人文理念，突出建设全国一流的、面向新世纪的大学，体现徐州的经济地位和地域文化特征。

行政楼位于南湖校区东入口处，对外交流办公便捷。建筑分设于入口轴线南北两侧，在南北入口前区设置广场，形成缓冲空间。本建筑位于校园主入口轴线的北侧，为校区内的领导及教职员工提供办公场所。在参照校方提供的用房布置建议的同时，将各类办公区域重新组合分区。校领导及院士的办公空间相对集中，并单独设置出入口，以确保清静幽雅的办公环境；其他教职员工的办公空间独立分区，出入口面向校前的广场，以满足办公需求。在首层将财务处单设办公大厅，方便直接对外。

本设计追求空间共享，用采光大厅以及吹拔空间处理方法将不同层次的办公人员联系起来，使空间渗透，凝聚团队力量。建筑内大部分房间采用南北向布局，充分考虑本地夏季气候炎热、日照充分的特点，减少建筑物西晒。利用室内外的绿化庭院形成建筑自身的生态气候，营造宜人的办公环境。利用柱廊产生庄重典雅的文化气氛，保持建筑风格的统一性，减少不必要的装饰，使建筑整体简洁大方。

IRVINE VALLEY COLLEGE PERFORMING ARTS CENTER
尔湾谷学院表演艺术中心

Architects : Arquitectonica
Partners-in-charge of Design: Bernardo Fort-Brescia, FAIA and Laurinda Spear, FAIA, ASLA, LEED AP
Location:Irvine, California,USA
Area : 5,446 m²
Client: South Orange County Community College and Irvine Valley College
Photographer: Paul Turang

设计机构：Arquitectonica
协助设计：Bernardo Fort–Brescia, FAIA and
Laurinda Spear, FAIA, ASLA, LEED AP
项目地点：美国加利福利亚州亚尔湾
面积：5 446 平方米
客户：南奥兰治社区学院、尔湾谷学院
摄影：Paul Turang

This is a 58,625 SF (5,446 m²) facility to house program components for three performing arts departments (drama, dance and music), bringing together students and faculty members for close exploration and collaboration. The facility's planning and design accommodates their individual needs as well as their common goals within a modest budget.

At the physical heart of the building is a 400-seat multi-purpose auditorium with a single level balcony that wraps around the orchestra seating, creating a intimate feeling. The proscenium stage with slight forestage projection and an orchestra pit capable of rising to stage level accommodates a variety of production and presentation types. The facility also includes a 99-seat black box experimental theater allowing for flexible staging and experimental productions.

The building takes advantage of the visibility of its site, being located at the edge of the campus and at the intersection of major vehicular thoroughfare, by making a bold architectural statement. In its plan and forms the building symbolizes the diversity and vitality of activities it houses. A folding angular wall wraps around the theater volume creating dramatic perspectives. This wall is also the organizing element around which other program elements are placed. The glass enclosed lobby allows light and view into the inner life of the facility while announcing the public entrance to the building.

East Elevation 东立面图

West Elevation 西立面图

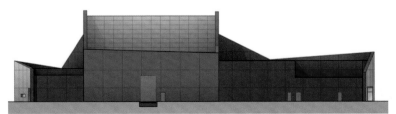

South Elevation 南立面图

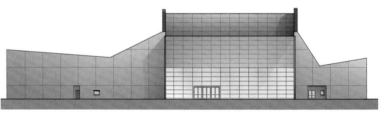

North Elevation 北立面图

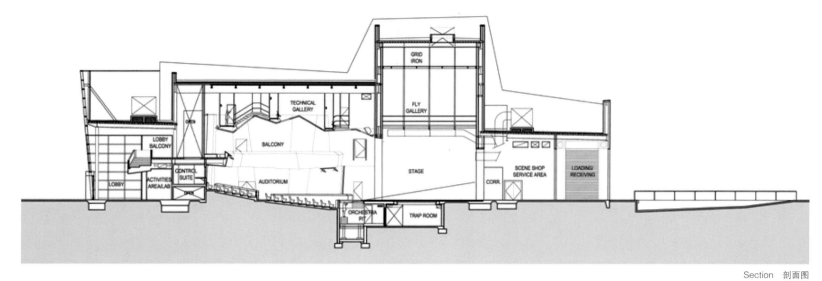

Section 剖面图

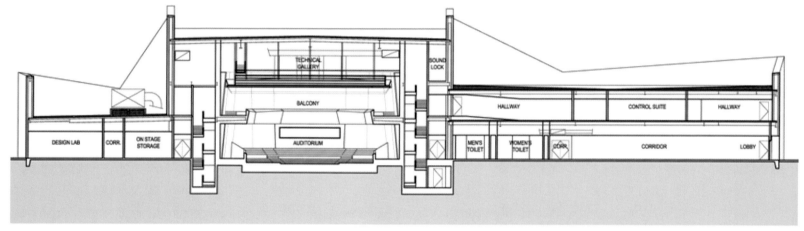

Section 剖面图

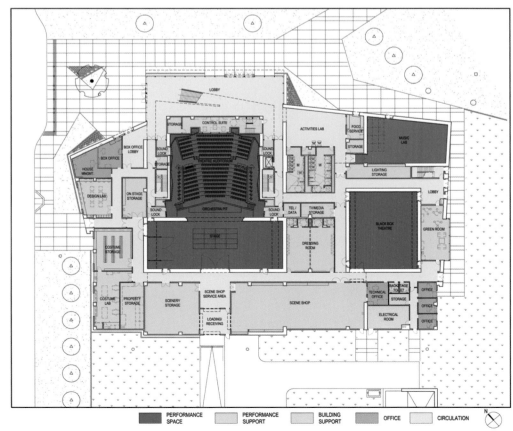

这是一个 58 625 ft²(5 446 平方米) 的建筑，由三个表演中心（戏剧中心、舞蹈中心和音乐中心）组成，汇集了紧密协作研究的学生和教员。建筑设施的规划和设计在有限的预算内满足了其需求和共同的目标。

在建筑的物理中心，有一个 400 个座位的带有单层楼厅的多功能礼堂，楼厅围绕着管弦乐队，创造出一种亲密的感觉。镜框式的舞台在幕布前方有轻微投影，且有一个可升至舞台的管弦乐队乐池，以适应各种表演和展示类型。这个设施还包括一个有 99 个座位的黑箱试验剧场，允许进行形式灵活的展演和演出试验作品。

建筑利用场地的可视性、处于校园的边缘和位于主要交通干道的十字路口等特点，进行了大胆的建筑学表达。建筑通过它的设计和形式表达了房屋的多样性和活力。影剧院体量被折角的墙壁包围，创造出一个引人注目的视角，这堵墙也是被其他元素围绕的组织元素。封闭玻璃大厅允许光通过，透过玻璃可以看到大厅内部的设施以及活动情况，建筑的公共入口也一目了然。

Plan 平面图

ZHEJIANG ART MUSEUM
浙江美术馆

Architects: China United Zhujing Architecture Design Co., Ltd.
Location: Hangzhou, China
Area: 31,550 m²

设计机构：中联筑境建筑设计有限公司
项目地点：中国杭州市
面积：31 550 平方米

The art museum is located beside the West Lake, at the foot of Yuhuangshan. The building is near the mountain and by the river, so it is richly endowed by nature. Its total construction area is 31,550 m², in which area underground is 15,338 m² and area above the ground is 16,212 m².

The building is developed along the mountain and falls level by level toward the lake surface. The building's contour line with fluctuation has reached a harmonious state between the building and natural environment. The building shape is natural and fully reflects style which is characteristic in Jiangnan culture. Color composition of pink wall and dark tile and modeling characteristic of sloping roof have been introduced into out work "between similarity and dissimilarity", which has reflected a refined and intangible cultural taste. A large area of white wall is set as the background, black roof component is used to draw the outline, its lines are free and strong, and it is full of aesthetic charm of traditional wash painting and calligraphy; application of steel, glass and stone has emphasized contrast of material quality; cones in different forms caused by roof deformation and alternate combination of horizontal masses have given a strong sculpture and modern sense. Unique architectural form and space fully reflect the profound combination of traditional charm and modern spirit, full of originality.

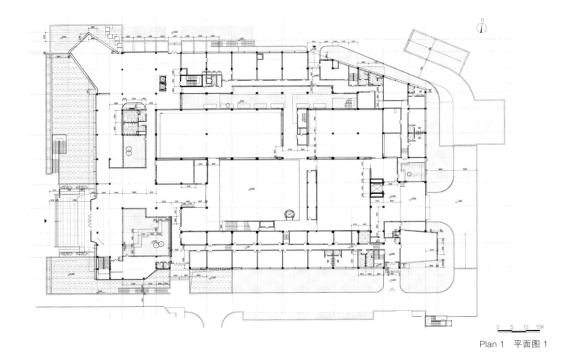

Plan 1　平面图 1

0 5 10 15M

美术馆位于西子湖畔，背靠苍翠的玉皇山麓，建筑依山傍水，环境得天独厚。其总建筑面积为 31 550 平方米，其中地下 15 338 平方米，地上 16 212 平方米。

建筑依山形展开，并向湖面层层跌落。起伏有致的建筑轮廓线达到了建筑与自然环境共生的和谐状态。建筑造型自然而又充分地流露出江南文化所特有的韵味。粉墙黛瓦的色彩构成、坡顶穿插的造型特征，在"似与不似之间"被带入了我们的创作之中，体现了一种清新脱俗而又空灵含蓄的文化品味。以大片白色墙面为图底，以黑色屋顶构件勾勒，线条张扬洒脱而又不失法度，极富传统水墨画和书法的审美韵味；钢、玻璃、石材的运用强调了材质的对比；坡顶变形生出多种形态的锥体与水平体块的穿插组合，使建筑充满强烈的雕塑感和现代感。

独特的建筑形体和空间充分体现了传统意韵与现代精神的深层次融合，具有原创性。

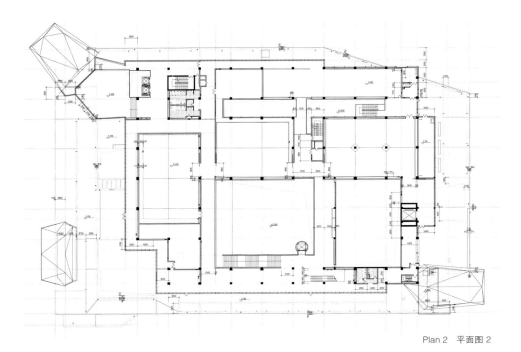

Plan 2　平面图 2

0 5 10 15M

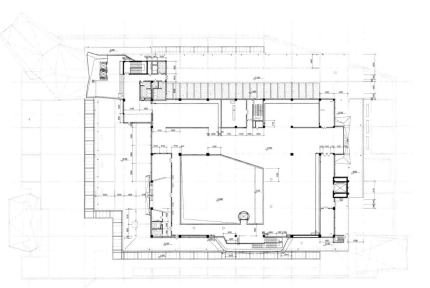

Plan 3　平面图 3

0 5 10 15M

CHINA ACADEMY OF SCIENCE & TECHNOLOGY DEVELOPMENT

中国科技开发院

Architects: CSC International
Location: Shenzhen, China
Area: 55 044.08 m²

设计机构：中建国际
项目地点：中国深圳市
面积：55 044.08 平方米

"Science & Technology Tower" is located in the east plot of South Park of Shenzhen Nanshan High-tech District, with TCL Tower at the opposite side separated by Gaoxin No. 1 Road to the north, South Gaoxin No. 3 Road to the south, and the completed office building and incubation building of the development academy to the west. This project is composed of a tower with a height of 100 m and an annex of three floors, supporting exhibition center and conference center of 6,000 m² (in the annex) and research and development rooms of 37,000 m² (in the tower).

South Park of Shenzhen Nanshan High-tech District is a landmark in high-tech areas. It is the obligation of every developer and architect to create and maintain benign urban space and building group style in the region, and the architectural space form should be seized via urban intention, which is the key point of general drawing design for "Science & Technology Tower". According to relations between "Science & Technology Tower" and the completed office building as well as incubation building in their masses, there are two modes on the general drawing which are "wrapped courtyard" and "open courtyard", and they are selected and analyzed from the angles of urban design and environmental benefits of the building.

塔楼立面比例分割　　　　　　　　　　　　　塔楼与群房地比例分割　　　　　　　　　　　塔楼以两个正方形拼接而成

Analysis　分析图

空间创新赋予建筑造型新意。"比例"是典雅端庄型建筑的关键，中置的空中庭院自然将塔楼分成双体量，比例经典，竖线条的处理使其更修长挺拔。东西立面相对较实，减少日晒，增添塔楼体积感，增加庄重气质。

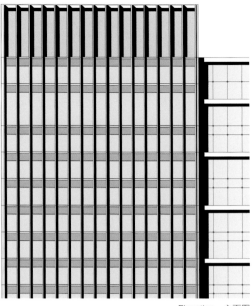

南北立面竖向修饰板使立面精致耐看，中部简洁平静的透明玻璃使内部的庭院环境充分展示，理性严谨的立面由于这一道立体空间的存在而生动。大厦整体开解比例成熟、简约庄重，充分展示项目的修改与气质。

"科技大厦"位于深圳市南山高新技术区南区的东片，属于深南大道南第二排建筑，北隔高新一路与TCL大厦相望，南为高新南三路，西为开发院已建的成办公楼与孵化楼院落。本项目由一栋百米塔楼与三层裙房组成，功能为6 000平方米的展览中心和会议中心（设在裙房），以及37 000万平方米的研发用房（布置在塔楼）。

深圳市南山高新技术区南区是高新技术区的地标，创造、维持区域内的良性城市空间及建筑群体形态是每个开发者及建筑师的义务，我们必须用城市意向把握建筑空间形式，这是"科技大厦"总图设计的关键。根据"科技大厦"与已建成的办公楼、孵化楼体量的关系，总图将出现"围合内院"和"开放庭院"两种模式，我们从城市设计、建筑环境效益等角度进行分析与选择。

NORTHUMBRIA UNIVERSITY CITY CAMPUS EAST
诺森比亚大学城东部

Architects: Atkins
Location: Newcastle, USA

设计机构：Atkins 公司
项目地点：美国纽卡斯尔市

Atkins' award winning Schools of Law, Business and Design were completed in August 2007 on time and on budget. The £47 m BREEAM excellent scheme consists of high numbers of cellular offices and teaching spaces, raked and flat floor lecture theatres and specialist workshops alongside cafes and ancillary accommodation. City Campus East was a major new build project on a brownfield site adjacent to a listed church in the centre of Newcastle.

Innovative solutions have delivered exceptional sustainability and meet the technical brief for space efficiency. This project places a key emphasis on: strong urban links to Newcastle city centre, creating pedestrian / metro friendly access, innovative structural design and means of construction, sustainability / environmental, flexibility within teaching spaces and how the language of architecture can exemplify the image of educational institutions.

Atkins was commissioned to provide full architectural services from RIBA Stage D onwards and was subsequently novated to the main contractor under a single stage design and build contract.

The building was recognised by CIBSE, winning the low carbon new build development of the year award. The specially designed solar shading facade, green travel plans, integrated renewable energy sources and building services strategy all contribute the BREEAM excellent rating. To minimise fabrication costs, the facade was designed so the angles of the tubes created a faceted repeating pattern. This meant that despite the seemingly complex shape of the spirals, only a handful of standard components were required.

The building is an eye-catching design, testament to Atkins' architecture and a brand building identity for the University of Newcastle, making a confident statement about ongoing regeneration and future of Newcastle. The building won Best Public Sector Award and Best Overall Building in the Northeast's regional landmark awards and has subsequently been shortlisted for the RIBA awards.

Modelling 1　模型图 1

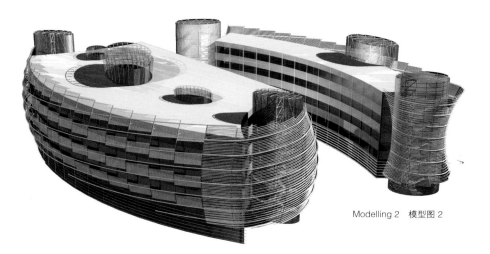

Modelling 2　模型图 2

　　Atkins 公司在法律、商业学院建筑的设计竞标中中标，建筑在规定的时间和预算内于 2007 年 8 月按时完成。这个造价四千七百万、按照建筑研究所环境评估法设计的优秀方案包括了大量的办公室和教学空间、阶梯教室、专家工作室以及咖啡室和辅助休息区。大学城东部是在毗邻纽卡斯尔市中心教堂的荒地上建立起的新建筑。

　　创新的建设方案传递了可持续发展的理念，并且在技术的支持下满足了空间的高效性。项目的设计理念重点强调：连接到纽卡斯尔城中心的强大城市脉络，创造行人、地铁友好的通道，创新的结构设计和建设方法，教学空间的可持续性、环境灵活性，怎样用建筑语言体现教育机构的形象。Atkins 公司被委托以英国皇家建筑师协会 D 级标准的身份提供全套的建筑服务，包括随后作为单体建筑设计和建造合同下的主要承包商。

　　本建筑是英国屋宇装备工程师学会认证的建筑，并且赢得了本年度的低碳新建筑发展奖。特别设计的太阳能遮荫系统、绿色旅行计划、集成再生能量资源和建筑服务策略都为本建筑的英国屋宇装备工程师学会卓越排名做出了贡献。

　　为了将建筑的成本降到最低，建筑采用了一种逐渐消隐的模式，通过管道的角度差创造一个具有小平面的重复形式。这意味着尽管建筑具有看似形状复杂的螺旋，但是只需要少数标准组件。本建筑是一个吸引眼球的设计，证明了 Atkins 公司的建筑品质。本建筑作为纽卡斯尔大学的品牌建筑，为纽卡斯尔正在进行的重建和未来的继续发展做了自信的陈述。本建筑获得了最佳公共部门奖和东北地区最具里程碑意义地标建筑奖，并入围了随后的英国皇家建筑师协会奖。

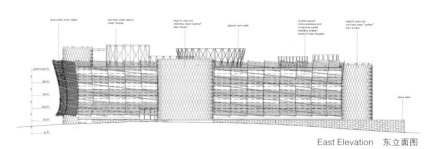

East Elevation 东立面图

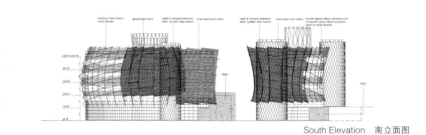

West Elevation 西立面图

South Elevation 南立面图

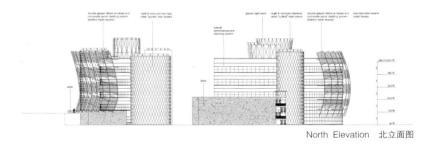

North Elevation 北立面图

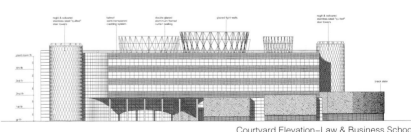

Courtyard Elevation–Law & Business School
法学院 & 商学院室外立面图

Courtyard Elevation–Design School
设计学院室外立面图

图书在版编目（CIP）数据

世界建筑事务所精粹：全3册 / 深圳市博远空间文化发展有限公司编 . — 天津：天津大学出版社，2013.5
 ISBN 978-7-5618-4622-3

 Ⅰ . ①世…　　Ⅱ . ①深…　　Ⅲ . ①建筑设计－作品集－世界
Ⅳ . ① TU206

中国版本图书馆 CIP 数据核字 (2013) 第 066708 号

责任编辑　郝永丽
策划编辑　刘谭春

世界建筑事务所精粹Ⅲ　　　深圳市博远空间文化发展有限公司　　编

出版发行　天津大学出版社
出 版 人　杨欢
地　　址　天津市卫津路 92 号天津大学内（邮编：300072）
电　　话　发行部 022-27403647
网　　址　publish.tju.edu.cn
印　　刷　深圳市彩美印刷有限公司
经　　销　全国各地新华书店
开　　本　245 mm×330 mm
印　　张　60
字　　数　810 千
版　　次　2013 年 5 月第 1 版
印　　次　2013 年 5 月第 1 次
定　　价　998.00 元（共 3 册）

（凡购本书，如有质量问题，请向我社发行部门联系调换）